DOLPHIN CONFIDENTIAL

Dolphin CONFIDENTIAL

Confessions of a Field Biologist

Maddalena Bearzi

The University of Chicago Press | Chicago & London

Maddalena Bearzi has studied the ecology and conservation of marine mammals and sea turtles for over twenty years. She is founder of the Los Angeles Dolphin Project in California and cofounder of the Ocean Conservation Society. She is coauthor of *Beautiful Minds: The Parallel Lives of Great Apes and Dolphins*. She lives in Los Angeles.

The University of Chicago Press, Chicago 60637
The University of Chicago Press, Ltd., London

Printed in the United States of America

21 20 19 18 17 16 15 14 13 12   1 2 3 4 5

ISBN-13: 978-0-226-04015-8 (cloth)
ISBN-10: 0-226-04015-1 (cloth)

Library of Congress Cataloging-in-Publication Data

Bearzi, Maddalena.
Dolphin confidential: confessions of a field biologist / Maddalena Bearzi.
p. cm.
Includes bibliographical references and index.
ISBN-13: 978-0-226-04015-8 (cloth: alkaline paper)
ISBN-10: 0-226-04015-1 (cloth: alkaline paper) 1. Dolphins. 2. Dolphins—Research. 3. Marine mammalogy—Fieldwork. I. Title.
QL737.C432B43 2012
599.53—dc23   2011030692

♾ This paper meets the requirements of ANSI/NISO Z39.48-1992 (Permanence of Paper).

To my mother and father, with love

Contents

Prologue

Dolphin Confidential follows an aspiration to share the story of my own experiences and passions, both in and away from the wild places where I have spent much of life in contact with nature. I hope to impart my deep love and appreciation for wildlife, recounting the moments of awe, the sensations I've had, and the wonder I've felt, along with the insecurities, doubts, and frustrations I encountered in the many years I've passed in the company of dolphins and other creatures.

This book includes many bits of scientific information as well as details of my research on marine mammals. I hope to provide readers with an idea of what a field marine biologist (or better still a cetologist) studies, and how I've gone about delving into the lives of dolphins and conducting scientific investigations in the wild. The central part of the book is devoted to the metropolitan dolphins of California, to which I owe most of whatever knowledge of dolphins I have managed to accumulate. I hope as well to shed light on the critical problems facing these magnificent, socially complex, large-brained, and emotional creatures.

This is also a story of how I have come to find my own way in the world and my own balance, sometimes against the design of the life I was born to, and sometimes in spite of obstacles that seemed insurmountable at the time. This story chronicles the transformative process by which I began my career in wide-eyed naïveté, then slowly shifted toward the belief that conservation and protection of nature is virtually all that matters.

The facts and encounters in these pages are real, as best my memory serves, colored somewhat by the course of time. The stories are told in a style reminiscent of a diary, moving chronologically forward (and at times backward) through my life as the months and years passed by. Some of the events have been compressed in time and space and, on occasion, adjusted to better convey the concept or experience to the reader. Some of the names of characters and organizations have also been changed to "protect the innocent."

In the last analysis, *Dolphin Confidential* is a journey into my mind as much as it is a journey into the vastness and magnificence of nature.

Acknowledgments

Nonhumans first. I am forever indebted and thankful to the animals I love for providing me a reason to delve deeper into the nature of their existence, and in doing so, to see deeper into my own. Amid the many creatures I have had the pleasure to encounter, a few stand out above the rest. They are the bottlenose dolphins of Southern California and my dog Burbank (who, sadly, no longer sits under my desk as I write these words . . .). As friends and companions, they have enriched every step of my life, endlessly teaching me lessons of empathy and compassion.

In the human world, I am immensely grateful to *Charlie*: husband, life companion, friend, adviser, even personal editor. My books would never see the light of day without you. With that said, there are no words that can ever express my most profound appreciation and unwavering love. *Mom* and *Dad*: for first walking me through nature and opening that magnificent window on the wild world, for always supporting my decisions and passions, and for dealing with my existence an ocean apart from you. *Gio*: for the formative scientific insights with this book and many others, and for sharing with me the same fascination for the ocean creatures we both came to be so intensely passionate about. *Marcy*: for untiring support, advice, and friendship. *My good friends and biologist colleagues*: Dr. Daniel Blumstein, Dr. Craig Stanford, Fabrizio Borsani, for their constructive role in patiently reading this book and exchanging opinions over the years; Dr. William "Bill" Hamner for his support with this book, for helping to shape my academic career, and for always believing in me; Dr. John Heyning for being an inspiration as a scientist, marine mammalogist, and all-around good human being; Dr. Bernd Würsig for thoroughly revising this book and inspiring me with his wonderful work with dolphins; and the anonymous reviewers for their constructive comments.

I am also thankful to the many wholehearted research assistants and volunteers who have worked long hours with me in the field and in the lab throughout the years, many of whom are now great friends. My thanks to Joanna Arlukiewicz, Karyn Jones, Paul Ahuja, Brigitte Steinmetz, Bettina Lynch, Silvia Paparello, Jason Chau, Andrea Bachman, Celia Barroso, Jennifer Bass, Andrea Cardinali, Mike Navarro, Shana Rapoport, Travis Davis, Jon Feenstra, Fumio Ogoshi, Alice Hwang, Daphne Osell, Mallory Mattox, Wes Merrill, Antonella Idi, Bob Arkwright, Lisa Openshaw, Monica Varallo, and all the others, too many to mention: but you all know who you are! I thank my old friends Manuel Mongini, Enrique Duhne, and Scott Eckert for their assistance with my research work, in both official and unofficial ways. I am especially grateful to Manuel and Enrique for helping me remember some names and facts lost over the years.

Of great support were all the environmental and research nonprofit organizations I worked for, including, among others, Europe Conservation, the Tethys Research Institute, and, close to my heart, the Ocean Conservation Society (which has funded my scientific research for the last fifteen years).

Thanks to Amy Krynak and Joann Hoy for their assistance through the editorial process; the great "book team" of the University of Chicago Press; and my editor Christie Henry for being always so enthusiastic and supportive about my writing.

PART 1

Magical Bonds

1

Yucatán

The pickup truck turns off the paved road toward the Reserva de la Biosfera de Ría Celestún, in the Yucatán Peninsula of Mexico. I watch out the window as the landscape becomes a warm desert of sand and spiny bushes. A couple of black vultures glide over a dead cottontail rabbit. My body has begun to lose that leftover humidity accrued over a long rainy winter spent at my home in northern Italy.

Behind me, a swirling cloud of dust slowly isolates the civilized world from my final destination, the remote research station of El Palmar. I was here one year ago: same dirt road, same truck jam-packed with ecovolunteers and provisions.

In the driver's seat, Eduardo struggles to keep the *camioneta* on the center of the narrow, bumpy track, trying hard to prevent the dense underbrush from scratching the paint off the side doors. Crammed in between us is a huge box of cereal, our breakfast for the next few weeks. The truck's old radio plays

a tedious Yucatecan song that fades into the background as my thoughts run free. I have a couple of hours to myself before being immersed, once again, in the microcosm of lectures to organize, equipment to prepare, questions to answer, meals to arrange, volunteer problems to solve, and night surveys to complete. I am thrilled to be here. I wouldn't care to be anywhere else in the world.

El Palmar is a place where "stuff" has no reason to exist. It's where a hammock under the stars is a million times better than a bed, where turtles, birds, and iguanas are companions, where the silence is so deep it seems unreal, where all windows open every morning to the vastness of the sea. It's where living in nature assumes its true meaning. It has nothing at all to do with where I am from.

The station is nothing much. An old lighthouse, a fisherman's shack, and a couple of outbuildings, shaded by a grove of coconut trees that grow in the only space not consumed by the dense undergrowth or the impenetrable mangrove swamps. A family of seven lives here, distant from any civilization. The children—one, two, five, seven, and nine years old—have never seen a world other than this. They walk without shoes, and dress in clothes passed down from their older siblings. Maria is the oldest. Last year, she would often sit next to me on the beach or near my hammock and ask me about towns and cities, and about what people do there. Our conversations were simple, limited by my beginner's Spanish, but it didn't matter. I once asked Maria about her family. She told me how her baby sister Lupe had died. Maria was five at the time, but she still remembers where Lupe was playing after an afternoon rain when she was bitten by a coral snake. A few hours later, Lupe was gone.

I think about the coral snakes I found near El Palmar last year. They are highly neurotoxic, causing a rapid death if no antivenin is administered. The king snake, a harmless mimic of its venomous cousin, also inhabits this area. They look almost identical except for the sequence of red, black, and yellow bands on their bodies. Before my departure, I memorized, "Red on yellow kill a fellow; red and black, venom lack." A victim of a coral snake bite must be hospitalized quickly, but being far away from the closest village and with no truck on site, none of us would be likely to make it to a hospital in time. I know that, and so do my volunteers. El Palmar was home to snakes, tarantulas, and scorpions long before it became ours.

Living here has its risks, but for any peril I could ever imagine, the rewards of staying at El Palmar far outweigh them. And it isn't just the beauty of the starry nights, the turtles, the quiet, and the sea; it is the sheer simplicity of life. Being a field biologist in this corner of lost paradise was my dream.

Eduardo taps me on my arm, snapping me out of my musing. He gently stops the truck and points to a gray fox standing in the middle of the road. The fox doesn't move and stands for a moment inspecting the truck full of people and supplies; then it turns casually away from the intruders and lopes off toward a Yucatán columnar cactus, one of the few still resisting extinction.

We are less than an hour from El Palmar, and the sky is darkening, preparing to storm. I yell to my crew in the open bed of the truck to cover provisions and equipment and put on their rain jackets. They comply promptly, full of excitement for the new adventure ahead.

It's pouring. Eduardo turns down the radio and focuses on keeping the truck on the road, now transformed into a slippery stream of mud. Returning to Celestún for more provisions won't be an option for at least a few days if the rain lasts any longer. Volunteers are quiet, trying to stay dry under a large plastic sheet.

After the deluge, an opalescent sky finally begins to clear, changing into the crimson shades of a staggering sunset. The thorn forest is emerald green in stark contrast to the wet sand turned bright orange by the abating sun. To our left, the mudflat is speckled with pink flamingos milling in the shallow waters in search of snails, brine shrimp, insect larvae, and algae. It's mostly algae and shrimp in their diet, high in carotenoid pigments, that give their feathers that pale reddish hue.

It's late in the evening by the time we reach El Palmar, the mud having made the journey somewhat longer than anticipated. A full moon illuminates the coconut palms and our way to the station. My volunteers are exhausted as they carry heavy bags full of clothes and sleeping bags toward their new home. They came a long way to experience nature firsthand and, though tired, seem in good spirits.

Back in Milan, I interviewed all applicants, as I needed to eliminate those who were not fit for this kind of hard fieldwork. I selected these ten out of twenty-three, and so far, I'm pleased with my selection. Some are older than I and are likely to wonder about my young age and my qualifications for being the principal investigator in this remote site. Tonight, however, they're too tired for questions.

Our research station is two plain cement structures adjacent to the fisherman's shack. One has a kitchen with an industrial propane stove and a large veranda with mosquito screens where we gather. The other is where the

volunteers sleep and where we shelter our gear from the daily tropical rainstorms. Outside, toward a wall of shrubs, there's a shower we built last year. My hammock hangs between two coconut palms near the beach.

The next morning I wake up early and rested. Miguel, our oversized cook from Mérida who arrived a couple of days ago, has begun to prepare breakfast for the crew. I plan the lecture and activities for the day and sort out field equipment for the coming night of research. This will be the first of many nights we will walk for miles along the beach, searching for nesting sea turtles.

From May to August, after venturing for years in the open ocean, hawksbill turtles come ashore at dark to lay their eggs in the white, fine sand of this yet unspoiled biosphere reserve. Their timing is faultless, as is their ability to locate the same remote beaches on which they were hatched, from hundreds, sometimes thousands of miles away. They do this by using exceptional orientation and navigation skills, "learning" the magnetic topography of their natal beaches. Fighting adverse currents, with poor sight and no landmarks, these vagabonds of the sea are able to determine their latitude and longitude and plot their migratory routes with a precision that would make any mariner green with envy. The ability to detect magnetic fields is something they are born with, an extra sense that comes in handy in the vastness of the oceans.

It takes one's breath away to see this ancient-looking reptile emerging from the waves under the light of the full moon. Though adept at sea, the terrestrial movements of a female turtle are quite demanding. Her front flippers leave deep tracks in the sand as she strains to pull her almost 135-kilogram body to the nearest slope. Away from the tidal ebb and flow that can flood her nest, she stops for a moment; a tear drops slowly from one of her eyes. She isn't crying, just purging the excess salt from her body. She modifies her land route slightly in search of the perfect nesting place and, once finding it, begins to dig. It will take her over an hour to mold a nest shaped like an oversized wine decanter; at least another half an hour to lay 140-plus delicate, Ping-Pong-ball-sized eggs.

I crawl silently next to the pregnant female, along with two members of my newly trained team. The others are busy tracing and measuring the fresh turtle tracks still visible on the sand. We've walked over eight miles along the shoreline tonight to find this female. Now, in the heat of this tropical Mexican evening, sweat and fatigue blend equally with the excitement of the scene.

Equipped with headlamps, paper, pencils, and measuring tools, we wait until the female lays her first run of eggs. Intrusive as it may seem, this is the best time to collect our data. The turtle doesn't notice our presence. Like an automaton, she lays a few eggs, pauses to breathe heavily, then lays more eggs until her job is done. Finally relieved of her burden, she returns to the sea, disappearing into the darkness of the waves.

As she leaves us, I wonder if she'll ever come back, if any of these ancient turtles will ever return here again. Threatened by shell trade, loss of feeding and nesting habitats, incidental mortality in fishing gear, pollution, and coastal development, the gentle hawksbill is just a few steps away from extinction.

It's 3:00 a.m., and we pick up our tools and strewn clothing from the moist beach sand. Together and tired, we begin the long walk back to the station.

I've been asleep for about three hours when a mature coconut drops near my hammock, waking me abruptly. From where I lie, I can see that my research team is still wrapped snugly in their sleeping bags, and I hear the drone of Miguel's heavy snoring from his hammock in the kitchen. I try to fall back asleep but can't.

The first rays of the sun warm my body as I sit on the beach, and the night's humidity seems to abate as the day begins. The silence is broken only by the noise of double-crested cormorants as their wings graze the water surface. A magnificent frigate bird soars over my head, lifted by the gentle breeze.

There is a bottlenose dolphin moving in my direction; it stops, inspects the bottom, and resurfaces with a large fish held firmly in its mouth. Its dorsal fin is deeply indented with a V-shaped notch. Right away I recognize Superhero, a dolphin I encountered here last year. Most of the investigations carried out by the Tethys Research Institute, for which my older brother, Giovanni, works, is dolphin research, so it was easy for me to get pointers from him on how to study dolphins.

Busy as I am now with sea turtles and volunteers, I have little spare time to study dolphins. But their presence here is hard to ignore, so I run back to my hammock to get my notepad and pencil. Without either of us knowing it, Superhero and I have opened the first chapter of my life with dolphins.

Two weeks have passed since the volunteers and I arrived at El Palmar. Our provisions are running out. It has been raining so hard for the last several days that the road is impassable, and a trip to town is out of the question. Another day like this and there won't be food enough for regular meals. We are cut off from civilization.

Persuading the fisherman to take two of us to Celestún aboard his *panga*, *Rosa*, is not hard, especially when I offer some money and to buy goods for his family in compensation. It will be a long ride at sea . . .

The rain is coming down hard as the fisherman, Miguel, and I push off the beach into the oncoming waves. After two hours at sea, the gray sky gives way to a stunning cobalt blue. The ocean is flat and clear as we pass near a school of twenty barracuda moving in a circle, feeding avidly on a shoal of mullet. Miguel ties a large bandanna over his bald, toasted head. I watch for dolphins with my binoculars and take notes on the weather and sea conditions. The fisherman doesn't say a word; he looks at the sea as if we didn't exist; then he sets a line from the boat and starts fishing. By the time we reach our destination, two groupers lie dead and bloody on the floorboards.

Celestún is alive with colors and music. We follow the warm smell of fresh tortillas and *carnitas* toward the main plaza, where the locals are gathering for a town fiesta. Under the cover of a royal poinciana tree, we find relief from the midday sun. There is only one market in town, and we walk there after a quick stop for tacos and Coca-Cola.

Rosa heads back out to sea; her profile is much lower now due to the combined weight of our bodies and the large boxes of food we've acquired. We made a good speed of six knots on the way here, but we are now making only three as the little boat struggles forward. We look at each other nervously as the wind picks up and the whitecaps intensify.

Twenty minutes later, waves are breaking over the bow, and the boat begins taking water in earnest. With a plastic scoop and a coffee mug, Miguel and I bail furiously to keep the ocean out while the fisherman works to maintain the boat on course. The floorboards are floating, and the dead groupers slosh back and forth around our feet. Cans, vegetables, pasta, orange juice, and bread find their way out of the soaked boxes. Two hours pass, and the water is still coming in as fast as we can deal with it. Our arms are sore from bailing, and we are awash with salt and sweat. Miguel pats me on the shoulder a couple of times in an attempt to cheer me up, but he knows I can't keep this up much longer. It is late afternoon by the time the ocean finally decides to be nice to us; the foamy waves begin to subside, and the residual water in the boat gently drains out of the stern scupper. A group—or school—of bottlenose dolphins joins us, but I am too tired to find my notebook, so I scribble some notes on the side of a wet box.

Some dolphins ride our bow wave, others glance at us occasionally while swimming near the boat. I count them, look at their behavior, and make a few notes as best I can. As I write, I notice something odd. The more I look at the dolphins and the more I write, the less exhausted I become. My energy is focused on the dolphin school, and I am completely absorbed in the moment.

After staring at me for a while as if I am some kind of freak, the fisherman finally breaks his long silence, asking me in a polite way, what the hell am I doing observing "fish" like this. I don't really know what to tell him. I explain that I have done this since childhood. First with cats, then dogs, hamsters, birds, toads, lizards, and snakes. Now it's dolphins and sea turtles. When I am fond of a creature, I tell him, I develop this need to explore its life.

I had animals around me from when I was five. It didn't matter if I lived in the city; there was always a place to find them, in the garden or the hills nearby. In the style of the early ethologists, I had my field book where I would write notes on everything I observed, including my dog's sleeping habits, a lizard feeding on a caterpillar, and the daily movements of the tortoise that lived in the yard. Reading Konrad Lorenz and Niko Tinbergen gave me early inspiration for my childhood annotations. I learned how to observe and how to be patient. By recording the number of actions and the amount of time spent by an animal performing a specific action, I learned how to create an ethogram, a detailed catalog of discrete and often stereotypic behaviors displayed by a species. The older I got, the more comprehensive my efforts became. I am trying to explain to the fisherman how I "collect data" on the dolphins around us. It probably doesn't make much sense to him in the end.

The dolphins are gone. Miguel is resting, the fisherman is back to looking at the sea in silence, but I can't stop thinking about what I might learn from these animals just from being here and watching them. The nights are already reserved for sea turtle studies, but I could use the morning before our daily lecture to have volunteers take turns observing dolphins from shore and recording what they see. I might even convince the fisherman to take us out once a week to follow dolphins and record their behavior. I have my camera with me, so I might also be able to take a few photos. I continue to ponder these new research ideas until my mental meandering is interrupted by the lighthouse of El Palmar coming into view. Some volunteers are sitting on the beach as we approach. Tonight we won't be looking for sea turtles: just a late dinner and a good and well-earned night of sleep.

The provisions surviving the boat disaster are devoured in two days' time. Appetite seems to grow exponentially at the field station in the last week; even Miguel's pineapple pastas and purple Jell-O are consumed at a frightening pace. It has to be the lengthy night walks coupled with the continual dips in the ocean to cool off. But nobody seems to mind much, especially as my pallid team of research volunteers have transformed into a group of healthy, dark-skinned hard bodies. It is something of which they are all quite proud, and they have taken to wearing almost nothing, most of the time.

The road to Celestún is finally dry, and we head back there again for food. As the coastal dune vegetation gives way to the prickly thickets of the inland, I miss El Palmar. Celestún's fiesta has passed, and the town is back to its quaint look of a deserted fishing village where residents keep cool in their hammocks or venture out for cold *cervezas* and television at the local Nicte-Ha restaurant.

We pass a group of vibrantly dressed tourists speaking broken Spanish. They are negotiating to hire guides to take them to the flamingo colonies along the river. Tourists here are a new thing. As I watch them board the *pangas*, I selfishly hope they will never learn of the pristine beaches of El Palmar.

At the market, I call to check in with my boss at Europe Conservation, the ecovolunteer organization I work for in Milan. He is shouting into the receiver as if to compensate for the distance between us. I imagine him sitting in his office, surrounded by stacks of papers and boxes of ecovolunteer T-shirts, in harmony with that hectic city life. He asks me if I am interested in being principal investigator for one of their marine mammal research projects in Greece as soon as I return from Mexico this summer. I am both shocked and excited as I stammer out an enthusiastic yes.

Miguel takes the turnoff for El Palmar one-handed while simultaneously stuffing the last remaining *pastelito* into his mouth. Sitting next to him in the truck, I think about the coming week of sea turtle research at the station, the dolphins of El Palmar, and those I will follow soon in the turquoise waters of the Ionian Sea.

2

Among Lizards

Back home in Padua, I've traded my hammock for a cushy bed for the first time in months. After Mexico, the sticky summer heat of the suburbs does not bother me as I peruse the long list of things I need to do for my coming adventure in the Ionian Sea. Already missing the quiet, the ancient-looking turtles, the simplicity of life that filled my days, I close my eyes for a moment and let my mind run free through the last several years of my life. How did I get here?

It seems only yesterday when I left this same room, charged with enthusiasm, to walk through the doors of the Biology Department at the University of Padua. It was my first year as a student of natural science: a degree, I thought, would open the way to a career as a behavioral ecologist. But school wasn't what I anticipated. The university lectures were far from engaging and had little to do with my idea of what "natural science" ought to be. It didn't take me long to feel suffocated, trapped inside a sterile classroom in the cement box of my university.

As a little girl, I remember looking out the back window of our family's green Simca at the swirling cloud of dust on the road to the remote campsite at Orvile, in Sardinia, where I spent many childhood summers on pristine beaches before a turquoise sea. Far away from my life in the city, it was there I first fell in love with nature; and it was there I would have stayed, if I could have. But I grew up in a succession of northern Italian cities, due to the itinerant nature of my father's job. The backdrop of my childhood wasn't the natural places I longed for. It was, instead, colored with the subdued hues of buildings that seemed to reproduce at breakneck pace in the ever-present winter fog of Veneto. In middle and high school, my experiences with nature took place, for the most part, through the dissected bodies of frogs, rather than outside in the real world where frogs might live and flourish. As an escape, I sought my own intimate "window on nature," by observing and collecting all sorts of small creatures, then inventing my own ethological experiments.

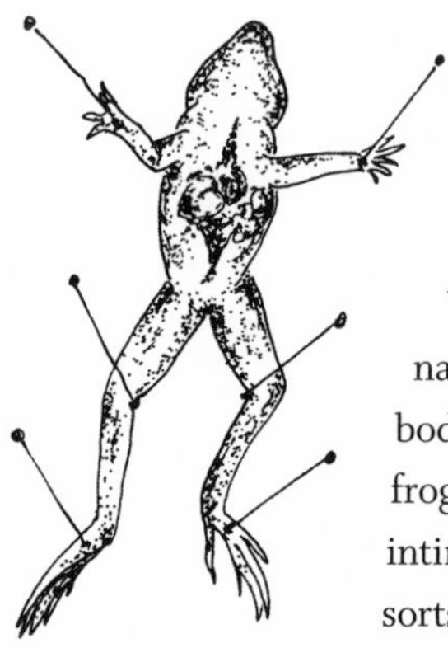

Moving to the university, I quickly realized that something needed to change, and I began looking beyond that concrete box for some bond to the animal world. I wasn't ready for the preprogrammed, provincial life that was unfolding before me. My options, which seemed to include marrying my high school sweetheart, buying an apartment in the suburbs near friends and relatives, raising a couple of kids, and working a nine-to-five teaching job in Padua, or just becoming a housewife, gave me goose bumps. Just talking about it gave me goose bumps. But everything was pulling me in that direction: where I lived, what my friends were doing, my boyfriend, the conservative Italian mentality, university. Breaking out of this mold would not be easy.

Looking for a way to make my university years mean something, I found a kind ally in a botany professor who spent most of his life cataloging plants in a spacious office on the second floor of the Orto Botanico di Padova, the world's oldest academic botanical garden. One cold day in February, we walked together past the 400-year-old palm that Goethe referred to in *The Metamorphosis of Plants*, past the historic greenhouse and into the Ortus Sphearicus, where we found a bench under a ginkgo tree from the mid-1700s.

"Believe me, Maddalena," he said, gently touching my shoulder, "I understand your passion for nature, but it will be hard to convince an adviser to take you for a field thesis on animal behavior in this department. Ethological studies are not really what we do here . . ."

"How about you?" I asked. "What if I find an external adviser from another university willing to accept an ethological thesis, will you be my adviser then?"

"Well, in that case we can talk," he replied, stroking his Santa Claus–like beard with his fingers.

When I was about twenty, I began working as a freelance photojournalist for a bicycle magazine in Rome. It was not much to speak of moneywise, but the job provided enough to pay for some school expenses and to cover the cost of occasional trips. Now it would pay for traveling around Italy in search of an external adviser interested in my animal behavior studies.

My job also gave me a means to leave the life of a city dweller for weeks at a time. The work entailed taking a bike trip to an interesting area, shooting pictures, and writing about it. My high school boyfriend, Marco, often collaborated with me on these assignments. They were our excuse to spend time together. Marco wasn't really the "adventurous" type—quite the contrary. His passions lay more with reading, photographing architecture, and working in his home darkroom. With his pale skin and round spectacles, he seemed more the intellectual sort. People say that opposites attract, and perhaps it was these fundamental differences between us that would keep us together for several years. I bicycled on the fringe of the Sahara desert and sipped mint tea in Tunis, camped in the Extremadura's Sierra Morena, pedaled hundreds of miles through remote areas of Italy's deep south and through most of the European continent.

Armed with some cash in my pocket from the bicycle travels, I knocked on many doors at universities all over my country, all of which were closed until finally, at the Ornithology Department of the University of Parma, Dr. Natale Emilio Baldaccini opened his. Explaining why I was there wasn't difficult, given the many times I'd rehearsed the speech in my mind.

At the end of my pitch, Dr. Baldaccini smiled and with a robust Tuscan accent said, "Sure, why not." Then, pointing to a hippie-looking guy wearing trekking shoes, he added, "That's Manuel, my right hand and research assistant. You can start by helping him with his investigations until you figure out what you want to study for your thesis." With that, Baldaccini walked off to his next lecture, leaving me alone in his office with Manuel.

From that moment on, when I wasn't taking mandatory courses in Padua or traveling for the bicycle magazine, I was with Manuel. He was my bridge to fieldwork and seemed to represent the outdoor life I was seeking. With a backpack on his doorstep, Manuel was always off to new adventures, placing feral pigeon traps in Austria, hiking the Monte Bianco, driving his rotten old camper through central Italy collecting reptile specimens, or off to Africa for months at a time. One could almost smell the scent of Africa, walking into his home on the shaky second floor of a century-old farmhouse in the Emilia-Romagna countryside; it was a place filled with incense, African artifacts, books, and music.

Manuel would call me spontaneously, saying, "Ciao, Maddalena, I just got back from Tanzania. Would you like to come along and tag birds in Sardinia tomorrow?" That was enough for me to pack and go. I too began living with a backpack on my doorstep, ready to leave at a moment's notice. Manuel was my guide toward a new, unexpected world of fascinating animals. He taught me how to be at ease in nature, how to appreciate the beauty of the most hideous or unexciting creatures. Driving down back roads with Manuel in Sardinia at 5:00 a.m. to tag birds, crawling along a riverbank in Emilia-Romagna waiting for sand martins to emerge from their nests, hiking through brush in search of foxes, or turning over rocks in Tuscany to count snakes, I began to feel like a field biologist. And I found a new home in the outdoors.

Sea turtles and dolphins hadn't yet entered my life, and studying the homing behavior of a lizard commonly found in central Italy seemed an acceptable compromise between the strict requirements of my university and the scientific interest of Dr. Baldaccini, who agreed to act as external adviser for my thesis. Baldaccini suggested that I apply his proven methodology for investigating pigeons' home range and homing behavior to a study of *Podarcis sicula* lizards—aka the Italian wall lizards—but how to practically accomplish this was another problem altogether, and mine alone to solve. The deal came with free lodging at the Tombolo Reserve, located in a protected pine forest in the Tuscan countryside near Pisa, where these lizards were abundant. My quarters were located in a large military-looking edifice, one-third of which was used to host students and university personnel from the University of Pisa; the rest of the building provided a home for caged pigeons and mice used in various experiments. From time to time, one of the white mice would get loose and join me for dinner in the large kitchen, glaring out from under the stove with red beady eyes.

My botany professor in Padua agreed to be my thesis adviser but not without a condition: for my thesis to be accepted at the University of Padua, I would need a botanical component: a hefty catalog of plant specimens present in different areas of the Veneto region. Essentially, I was signing up to produce two theses in one . . .

With my Fiat 500 filled to the roof with bags and research instruments, I left home to spend that summer—and the following ones—studying lizards. Even though Tombolo wasn't the wildest place in the world and lizards certainly weren't the most intriguing animals one could ever dream of, I was outside, in the field, studying animal behavior on my own as I had always wanted.

What followed were months often spent in isolation from my own species, working on my weird experiments. I refined the best way to capture lizards with a slipknot, measured their internal temperature, and marked a number on their backs with an indelible pen. Then I observed both the home range and behavior of the marked individuals from sunrise to sunset. Finally, I designed a structure for my homing trials in an open grass field located near the lizards' dwellings.

Anyone peering over the fence of the reserve during my last season at Tombolo would have seen a tan girl in shorts and bikini top, standing on top of a six-foot-high wood platform, plucking lizards from a bag and dropping them, one by one, through a hole located in the center of the platform. All around the wood structure, I'd marked the ground with concentric circles at a fixed distance, divided into numbered sectors. Viewed from above, it would have looked like a giant grass dartboard. Those same concentric circles and sectors, together with a topographic aerial map, were drawn on a clipboard.

The thought behind all of this was to release one lizard at a time from the platform, letting it hit the ground with random orientation so I could determine how long it took each individual to find the direction toward its home. Then, following the movement of each lizard from the release point with binoculars, I would record its trajectory on the map. The same experiment was conducted at increasing distances from the lizards' presumed homes. At the end of each experiment, I would spend weeks checking whether marked and released individuals actually returned to their original "home," thereby determining their homing ability.

The results were remarkable. My tiny roving reptiles were not only able to detect their homes over what was for them the massive distance of a few hundred meters, but a statistically significant number of them could find their way back from the release point, demonstrating an outstanding homing ability. The day I presented my thesis, "Home Range and Homing of *Podarcis sicula campestris* (Reptilia, Lacertidae) in the Tombolo Reserve (Pisa, Italy)," I managed to impress my stoic advisory committee at the prestigious and somewhat old-fashioned University of Padua.

The leap from lizards to sea turtles came naturally. At the time, my brother, Giovanni, was studying dolphins in Croatia with Tethys, a Milan-based research institute founded by Giuseppe Notarbartolo di Sciara and set up to investigate and protect marine mammals and their environment. Tethys was conducting research with the support of ecovolunteers from a sister organization, Europe Conservation. The formula of having volunteers pay for firsthand experiences in the company of cetaceans was the economic engine that kept Tethys's field research running. The money collected over the course of the summer season not only covered the full cost of the research but helped pay some of the annual administrative expenses as well.

Europe Conservation, whose focus extended beyond marine mammal research, was looking for someone interested in developing a brand-new ecovolunteer sea turtle program in a remote Yucatán reserve. Somehow, studying reptiles (albeit of an utterly different taxonomic order) and my willingness to stay in an isolated location for months on end got me hired. Thrilled as I was, I didn't let the fact that I spoke no Spanish, or that the job paid next to nothing, get in my way.

It's odd how the current of life can take you unexpectedly from one place to the next, sometimes without even a good grasp of what just happened.

3

Departure

As I look out the window from my passenger seat, my parents are waving good-bye before turning into black dots on the dirty glass of the train. I push my backpack onto an upper shelf, straining to make room. The compartment is full and sweltering, the air heavy with smoke. A couple of military guys are eyeing the breasts of the young woman seated across from them. They are trying to be nonchalant, but it isn't working and she disapprovingly adjusts her blouse. I pass unnoticed: jeans, T-shirt, tennis shoes, and, on my lap, my Moleskine Bruce Chatwin–style, and a few books. As we leave the ash-colored buildings of Padua behind, the train cuts through the agricultural flatlands that border the Po River. I settle back, envisioning the weeks ahead, relaxed by the monotonous sound of the railroad.

I am on my way to the southern Italian port of Otranto, on the Salento Peninsula of Puglia, where the research vessel *De Bolina* is docked and ready to cross the Ionian Sea for the Greek islands. I have been hired as the principal investigator for marine mammal research aboard the *De Bolina*. I feel a growing tension about this new and unfamiliar task as I begin to mull over the specifics. Field experience and dealing with volunteers are not an issue, considering my work with sea turtles in Mexico, but what do I know about dolphins other than some peripheral reading and those brief encounters with the bottlenose back in El Palmar?

In the few weeks leading up to my departure, I devoured stacks of marine mammal scientific literature, field guides, and books. I've learned much about cetacean ecology, organized lectures for the volunteers, set up basic methodology for fieldwork, memorized names and sketches of dolphins and whales to better recognize the various species we'll encounter once we're at sea. My lack of experience with marine mammals is offset by a head full of theoretical knowledge, but, of course, something is still missing. As my thoughts meander back and forth over what I know and what I don't, my eyes grow heavy and I fall asleep.

The shrill chime of the bell announcing the arrival of the food trolley wakes me up. The landscape has changed from the colorless fields of northeastern Italy to scenic hills where, from time to time, rows of cypress trees guide the way to a *casale*, a sort of cross between a grand villa and a farmhouse. My *Sierra Club Handbook of Whales and Dolphins* by Leatherwood and Reeves has slipped between the seats. I fish it out and open its cover. There, on the first page, is the inscription "Che accompagni bene te come ha fatto con me nella nuova ventura! Fabrizio." [May this keep good company for you, as it did for me, in your new venture! Fabrizio.]

When I first met Fabrizio, he was a young bioacoustician aboard a dolphin research cruise in the Tyrrhenian Sea, where I was tagging along on the sailboat as a friend of Manuel. Over the years that followed, Fabrizio and I became good friends, but I remember little of him on that cruise. Unfortunately, my most vivid memory was spending the entire trip belowdecks, seasick as a dog. My romantic ideas of sailing free into the wind and learning the ways of a dolphin field researcher had brutally clashed with the reality of being green and nauseous. And the worst thing was that I could not do a thing. I could barely get out of my bunk. Now I was beginning a new venture as a dolphin biologist in Greece . . . aboard another sailboat.

Trying not to think about seasickness, I nervously thumb through the pages of my guide. Hundreds of miles, a couple of train changes, one book, and a nap later, and the train speaker finally announces, "Staaaaaaaazione di Otranto." I have arrived.

I stand outside the station under a late afternoon sun in the heel of Italy. The summer heat is relieved by a light sea breeze, and there is an inviting smell of

panzerotti, a fried calzone "*pugliese* style." A taxi is too expensive for my pocketbook, so I shoulder my heavy backpack and walk toward the harbor, located just outside the fortified walls of the city. It's a long stroll, but I am so excited I barely feel the distance.

The small harbor, one of the main eastern ports of the Roman Empire, now shelters a crowd of brightly painted fishing boats rocking on the turquoise waters of the Gulf of Otranto. The skipper of the *De Bolina* is waiting for me, sitting on deck of his fifty-one-foot ketch. He has the stereotypic appearance of a sea captain: dark and skinny like a licorice stick, long salt-and-pepper hair, and a full beard that covers most of his oval face. Deep lines that bespeak a lifetime under the sun engrave his face in every direction like a spiderweb. He smokes a pipe, swinging it right to left, as he puffs. He's barefoot and wears dirty-white pants rolled up over his ankles, and a T-shirt with the sleeves cut off, which shows off his sun-tanned round biceps. As I approach, he scans me head to toe as though I were in an MRI machine.

"I am Enzo," he says with a broad smile and a heavy Venetian inflection. "Nice to meet you."

He sounds charming, seeming even more so when he bows to kiss my hand and invites me aboard. We exchange a few words, each trying to figure out the other; then he points at my shoes with a quick sign to take them off before boarding.

Each summer, Tethys hires Enzo and his boat to provide the research platform for dolphin surveys in the Mediterranean, staffed by ecovolunteers under the guidance of live-aboard young biologists from the institute. This is the first investigation of its kind in the Ionian waters, and this time, the researcher is me.

Everything on the *De Bolina* has a specific place, and the skipper insists his boat be kept in perfect order from stem to stern. Enzo explains this as he shows me to my cabin on the port side of the boat, my new home for the summer. My cabin is tiny, with two coffin-looking bunks, one above the other, and a small closet to hang my clothes. I take the top bunk. It has the only porthole in the room, providing something of a view.

The accommodations on the *De Bolina* are simple but comfortable enough. There are four open bunks in the forepeak and another stateroom to starboard, slightly larger than mine and with a double bed. Just aft of that, there is a small head with sink that I will share with the volunteers. The skipper and his mate

Jacopo, who's arriving tomorrow, share a large stateroom aft, complete with their own head and shower. They are isolated—"boatly" speaking—from the rest of us by the main salon and galley area, which houses a good-sized dining table with a built-in bench seating at least ten. Enzo makes it clear to me that "his" part of the boat is private, and no volunteers, nor I for that matter, are ever allowed to enter.

After settling in, I join Enzo on deck. He is smoking his pipe and sneering audibly at a couple of partygoers, moored several boats away from ours. He has already eaten dinner and smells faintly of *spaghetti alla carbonara*.

"Would you like a cold coca?" he asks me, opening a can and offering it to me. A late afternoon breeze blows over us from the land. We sit quietly next to each other, watching the activities of the port. He breaks the silence, pointing out some features of the harbor and telling me about his boat, which he built with his own hands.

We talk for over an hour, before Enzo asks me what I did before coming here. I begin telling him of my work in Mexico, and he launches into a narration of his past journeys at sea. I am so fascinated by his persona, I barely notice he has just cut me off in midsentence. I don't really care; there is plenty of time, and we'll be in the same boat for weeks . . . His stories are captivating. He is mysterious and interesting, unlike the university boys I usually hang out with in Padua.

The sun is going down, and Enzo is tired. "It's time for bed," he says. He leans forward and kisses me on the cheek. "Good night, Maddalena," he says, and disappears down the companionway. I am left alone on deck, flushed and excited by my first encounter with this strange, intriguing man.

"He's great and this is going to be a fantastic adventure!" I tell myself, while walking off the boat for a stroll. The city is beautiful, with that particular smell of the Salento Peninsula I remember from childhood vacations with my family along this same coast. The whitewashed old buildings of Otranto now glow reddish in the warm colors of a dying sunset, and people begin to flow onto the cooling streets, which crisscross under the watchful eye of the Aragonese castle atop the hill. Groups of middle-aged women chat in doorways, dressed in the funereal black of generations. Their husbands are probably at the local bar or playing bocce ball near the docks, awaiting the return of the fishing fleet.

Several tourists saunter along the historic stone pavement, peeking at the stores that stay open late into the night.

By 7:00 a.m. Enzo is awake, hands busy with coffee mug and pipe. He looks distant and pensive. It's clear to me he wants to be left alone.

"Maybe we should talk later about the trip and the research, OK?" I say, thinking he may have woken up on the wrong side of the bed. No answer. Heading to town for a cappuccino and a quiet place to review my research protocol seems like a good choice. The volunteers will arrive tonight, and I am still not certain whether my field methodology is the most appropriate for the coastal and offshore waters of the Ionian Sea and the species I expect to encounter there. I have designed a research protocol for collecting data on the animals we intend to study, based on reading and studies I have done, but it has yet to be field tested.

Our study seeks to gather preliminary data on animal distribution and frequency along with some information on their behavior and sea conditions. Among the things researchers struggle with when studying animals in the wild is finding the most appropriate observational techniques and systematic sampling procedures to use for their respective subjects. This is particularly hard when dealing with long-lived, free-ranging species like dolphins, which spend most of their lives underwater. Our scientific approach for this investigation would be longitudinal rather than cross-sectional, meaning that we would conduct surveys and collect data (samples) on a group of individuals, over whatever period of time we were able to follow them. A group, in our case, included all individuals in a school displaying the same "coordinated" activity, like traveling in the same direction or feeding. Our observational method would be what is called a "group follow" because we intended to focus on the entire group, not just individuals. Designing a quantitative sampling approach added yet another layer of complexity to the methodology; in theory I needed to decide how long to stay with a group, but because this was really a pilot effort, spending time with each group as long as possible sounded like a good initial plan. I decided we'd collect data on simple forms, open-ended enough to allow modification for unforeseen occurrences specific to the areas we planned to visit, and at the same time, complete enough to gather all the needed information.

Back in Padua, I talked methodology with my brother, Giovanni. On dolphin matters, Gio knows far more than I do. For him, becoming a dolphin biologist was a natural calling. As for me, I always showed a strong interest in science, but there were other things too, like working as a photojournalist and taking every opportunity to travel.

My father used to tell me I needed to make up my mind about what I wanted to do in life, and for a few years, I felt like I wasn't accomplishing much because I couldn't commit to a single career path. Anyway, to our family and friends, Gio was always the "dolphin expert," and I, somewhere down the food chain, was the reluctant and less important "reptile enthusiast."

I struggle to suppress another ripple of panic that sweeps over me as I finish my cappuccino and begin reviewing my notes in earnest. Not only must I teach volunteers something I have no experience with, I need to authoritatively collect data and resist seasickness at the same time. And I must do all of this in the shadow of my competent and committed brother, the dolphin expert.

That afternoon, I talk with Enzo about my plans for fieldwork, and we discuss the *De Bolina* itinerary once at sea. His mood seems improved, and he is back to the charming man I met yesterday. We plan to sail across the Strait of Otranto, between Italy and Albania, for the island of Corfu, then on to Lefkada, Kefallonía, and perhaps even Ithaca, if time and weather permit. We need to be back in Otranto at the end of our two-week "turn," to drop off our volunteers and pick up the next group. We want to invest as much time as possible searching for cetaceans but will seek out safe ports every afternoon so the crew can get a taste of these incredible Greek islands, filled with so much mythology and history.

Jacopo, our first mate, arrives midafternoon. Jacopo is in his late twenties, clean-shaven, with dark closely cropped hair in almost military style. He has a plain but likable appearance and looks very north Italian. At first meeting, I think he must not like me, as he speaks only to Enzo, never addressing me directly after our initial greeting. But I soon learn Jacopo is just a quiet type. He is aboard the *De Bolina* to make money and gain experience on the water to fulfill his dream of buying a boat and spending the rest of his life sailing off for destinations unknown.

Shortly before dinner, Enzo knocks on the door of my cabin.

"Follow me," he says, as he gently takes my arm, escorting me into the galley. "I thought you would like to see how I make *spaghetti alla carbonara*, so when we are at sea, you can show the volunteers how to make it. The eggs have to go in at the end, so they don't cook . . . like this . . . very, very important," he says, demonstrating.

On these research cruises, volunteers take turns cooking, washing dishes, and keeping the boat clean. Enzo tells me the skipper does not cook or clean once the cruise begins.

Enzo and Jacopo move out on deck to await the arrival of the volunteers. I remain below for a few minutes to make coffee, then climb the companionway ladder to join them on deck. Enzo does not see me and forges on with his conversation about what the new crew of *dolphinettes*, as he calls them, might look like. "I hope the girls are cute this time . . . We had some ugly ones last turn . . ." he tells Jacopo, who doesn't seem to be listening.

I clear my throat to announce my presence, and Enzo changes the subject and smiles at me, not knowing whether I overheard his musings or not. I return his smile without comment. The three of us sit quietly, listening to the evening breeze play the rigging like a lyre. The volunteers are late. Enzo smokes and fidgets.

We are expecting only four volunteer assistants for this first turn, three girls and a boy. They are two hours late when we finally see them wandering along the seawall. They are pale and tired from toting their heavy backpacks, and even though they are all Italians, they have the lost expression of foreign tourists. Putting two fingers in his mouth, Enzo whistles loudly, turning heads. They see us and excitedly jog toward the *De Bolina*. Their exasperation has morphed into plain tiredness, so we decide to keep introductions to a minimum and show them to their staterooms.

Everybody is up early, ready to begin the new journey at sea. Enzo introduces the "staff," pointing at Jacopo and me, highlighting our strengths and qualities with the warm tone of somebody who has known us for years.

This morning, he has finally changed the T-shirt he's been wearing for the last three days, and he's back in his amusing captain persona. He dotes over the three female volunteers, in an attempt to captivate them with his charm as he did initially with me. Then he becomes serious and professional, listing the many "don'ts," along with a few "dos," that govern volunteer conduct. In under an hour, Enzo gives a rapid-fire account of safety aboard, and a lesson on nautical terminology, sailing basics, and knots; discusses what to do in an emergency; and provides instructions on how to use the head, galley, and deck shower and other tidbits pertaining to life on the *De Bolina*. At the end of his protracted speech, Enzo lights his pipe and says, "I don't like to repeat things so I hope everyone understands everything." He flashes a last smile to the girls, makes a clicking noise, and walks away.

Four confused faces now turn to me. We all sit on deck and I explain that, until now, very little is known about cetaceans along the western shore of Greece. "There is a 1965 report mentioning the presence of fin whales, orcas, bottlenose dolphins, common dolphins, and some others in these waters, but their occurrence has never been assessed with rigorous scientific method," I begin telling them, "and this is our main goal: to investigate the distribution and frequency of all cetacean species in this stretch of the Ionian Sea. The truth is, we don't know much about cetaceans in this area, or in many areas for that matter. You can find lots of folklore about the relationship between humans and cetaceans, going back to ancient times, but if you look at what scientific data exists, only little was known until forty years ago." Enzo cuts me off, inquiring to no one in particular about his lost sunglasses.

"Somebody moved them from here," he says angrily, pointing at the hatch.

"So, I was saying," I continue, deciding to ignore him, "that cetology is a relatively new science, and only in the early 1970s, less than twenty years ago, scientists abandoned studies in captivity in favor of exploring free-ranging cetaceans. And that is challenging work; wild dolphins and whales are elusive in nature, and it's difficult to study them in open waters for many reasons. Even with new and improved techniques, we're just scratching the surface of what there is to know about them. That's why finding out what animals are commonly living here can give us a good starting point for understanding more about these creatures and their environment. You should feel like pioneers in the world of cetacean research!" When I finish my talk, I invite the crew to tell me something of themselves.

Nella is the first to introduce herself. She is near forty, with a welcoming face and long dark hair that she wears in a thick braid. She has volunteered in diverse ecoprograms around the world, often going to remote sites difficult to reach even by foot. Looking at her, I wonder how she could have done it all: Nella has a limb-length discrepancy; one of her legs is significantly shorter than the other, which makes it challenging for her to walk normally. She exudes a great enthusiasm for her past experiences as well as her expectations aboard the *De Bolina*. I like her immediately.

The other two girls, Mara and Lisa, are both in their early twenties, students from the Biology Department at the university in Milan. They came here to unwind from a hard semester of study and see dolphins up close for the first time. Mara is pretty and blond, with a turned-up nose and a thin, pale body. Lisa is her opposite, short and round, with dark hair and a wide smile that reveals a shiny steel wire crossing her upper teeth. They seem an inseparable unit, talking endlessly and doing everything together, perhaps to reassure one another.

Matteo, our token male crew member, is eighteen years old with an acne-inflamed face and thick glasses that make his eyes look like pinholes. He is tall and thin. Like a typical teenager trying to find his place in the world, Matteo is nervous and insecure in his demeanor. He doesn't seem to have much interest in anything except being wired to the earphones of his CD player, listening to Pink Floyd. It was his parents' idea to send him on this trip, and listening to him talk, it seems more like a punishment than a vacation gift.

We have decided to leave in the early afternoon. The day is stunning, and the Pugliese sun is hot and dry. The ocean is clear and calm, and there is no sign of scirocco, the warm, sometimes angry wind that blows from the Sahara desert over this region. All is now ready to depart for the Greek islands.

The engine is idling, the skipper is at the wheel, Jacopo casts off the last line holding us to the dock, and we are at last under way. As we pass the Otranto breakwater, Jacopo at the mast furiously grinds a winch, slowly raising the mainsail.

The old port disappears behind us as the *De Bolina* heads out into the open waters of the Strait of Otranto on an easterly heading. There are no questions now; we sit on deck, immersed in our thoughts and the new sensation of being at sea. The Italian coastline gradually loses definition, becoming a thin line on the horizon that slowly disappears as the afternoon wears on. A gentle sea breeze has come up, and Jacopo raises the jib and mizzen sails and turns off the engine. We are sailing, and the sound of waves gently caressing the bow replaces the last noise of the engine and of land. Our journey has started.

The deck of the *De Bolina* is comfortable, and the small but efficient cockpit is located far aft. Just forward of the mizzenmast, the main cabin hatchway leads down to the main salon area. The cabin top is slightly raised, leading forward to the mainmast, where it merges into the wide expanse of the foredeck, which is broken only by a large anchor windlass. The top of the cabin offers good seating for lectures and dolphin surveys, all in the shadow of the mainsail to provide some relief from the sun. The jib is raised on a forestay that runs from the mast to a bowsprit extending several feet over the water. Beneath this is a net for the crew to walk on when raising or lowering the jib. In calm seas, this is a spectacular place for close-up sightings of dolphins riding our bow wave or, in their absence, to rest suspended over the water, open to an unobstructed view of the expansive sky.

It's time for my first training lecture, and I've almost memorized the entire talk word for word, being so nervous about making mistakes. I gather my audience on the bow and begin teaching them about marine mammal ecology and how we will collect data once we sight cetaceans. They all seem quite receptive and interested.

"One of your main tasks is to search for dolphins and whales at sea from sunrise to sunset," I explain to them. "We will coordinate shifts so at least two of you are always on deck looking at either side of the boat, continuously scanning the surface and alternating between naked eye and binoculars. We will do this until it gets dark or ocean conditions worsen. If that happens, we'll stop searching for animals because they are too difficult to spot in waves."

I give them each a copy of the simple data collection sheet and explain how to use it when dolphins or whales are sighted. Date, time, latitude and longitude of sighting, weather conditions, species, group size, direction and movement, water depth, and distance from the nearest coast are the key data we will collect. I tell them they'll need to count the number of individuals every few minutes to check for changes in group size. At the end of the lecture, I feel slightly more confident in my new role of cetologist.

The sky is turning purple as my crew ends their first hours as dolphin researchers. As the last sunlight leaves us and the nighttime humidity begins seeping through our T-shirts and into our bones, we decide to regroup below.

"When are we going to see dolphins?" Matteo asks me, pulling a sweater over his head.

"I think the last team observed them on the crossing to the islands, is that right, Enzo?" I query.

Enzo is seated at the dining table doing something and limits himself to an "uh-huh."

Nella and I are sharing my stateroom, and the cooking duties for our first day at sea fall upon us. Nella, like a good Emiliana,* is an excellent cook and suggests we make lasagna for dinner. Enzo hears this and, looking up, says he rather feels a good carbonara is more appropriate.

"We've had carbonara since I got here, Enzo . . . Do you ever eat anything else?" I quip jokingly.

* A person coming from Emilia-Romagna, a region renowned for its outstanding cuisine.

"I like carbonara," he says without humor, and goes back to what he was doing.

In less than an hour, the seven of us sit down to dinner. The table is set with nautical-themed plates and glasses, a big loaf of bread, and a bottle of wine. Enzo ties a bib around his neck and launches into stories of the sea, alternating forks of *spaghetti alla carbonara* with gulps of wine. The girls are captivated, even more than I was my first day in Otranto. He tells the same stories I've already heard. To me, his "wolf of the sea" persona is beginning to break down one piece at a time, and I get up and begin washing the dishes.

By 11:00 p.m. volunteers and skipper are all asleep. Jacopo is taking his turn at the wheel while the *De Bolina* motors through still waters, the wind having deserted us with the setting sun. I sit next to him and stare at a moonless sky filled with stars; I am at ease for the first time in weeks.

Jacopo breaks the silence. He asks me about my work with sea turtles, and we start talking about Yucatán and traveling and places both of us would like to see. As we speak, a thread of common interest begins to unfold. Over the next weeks at sea, these evening talks would become a regular "date." Sometimes words would flow between us like an unremitting storm; conversation was easy and spontaneous. Other times we'd just sit silently, watching falling stars streak across the sky and disappear into the darkness of the horizon, or we'd stare intently as the water surface glowed and flashed with the greenish bioluminescence of marine organisms.

At sea, I felt like when I was in Yucatán: full of energy with little need of sleep. Any trepidation about seasickness was dissipating with each day that passed. Getting up early allowed me to work on lectures and resolve practical research issues while my crew was still in bed, and staying up late provided the time to sit and talk with Jacopo.

Our first landfall and taste of the Greek islands is Kérkyra, known as Corfu, the second largest of the Ionian Islands. After a two-day passage, we tie up in the main port and decide to stop for a day. We meander around town, letting our bodies get rid of the rolling motion of the waves, buying souvenirs, exploring the old fortress, the chapels, and brick streets, all filled with brightly dressed tourists.

"So, do you think we'll see dolphins tomorrow?" Matteo asks me on our way back to the dock.

"Well . . . I can't tell you for sure, but we will do our best to find them." What else can I say?

Back aboard, I argue with Enzo about some plastic bags Mara left on the chart table. “Can’t they leave the boat in order? It’s not their goddamned house here!” he snaps, flinging the bags across the cabin.

The next morning we have an early wake-up call and a quick breakfast before departing Corfu. Then it’s everybody on deck, ready to search for dolphins as the sun peeps over the horizon. The next few days are filled with new things and places to see and learn. We travel south to Paxos, with little wind, then on along the mainland coast to Lefkada. And although we are vigilant in our search, we find no dolphins.

One day we see a dead loggerhead turtle bobbing like a log in open water. Another day, to our amazement, a single swordfish leaps high above the surface, landing with a spectacular splash. But aside from some of these isolated encounters, we see little else. We arrive as far south as the port of Fiskardo on Kefallonía before we have to turn back northward. The weather has been exceptional, but the winds are light, and we motor much of the time. Sometimes we spend late afternoons swimming and sightseeing in one port or another.

As we head back, the only thing missing from our journey is dolphins. The close encounters with cetaceans that my volunteers signed on for seem further and further from reality. Matteo asks me every day whether we will see dolphins, and recently Mara and Lisa have taken to asking similar questions. “Are you sure we are going to see them before returning to Otranto? There are not so many days left . . .”

“Unfortunately, I can’t predict where they’ll be,” I tell them. I am beginning to feel as if I’m letting the volunteers down. After all, what do I really know? I’ve never sailed here before and have only seen dolphins a few times in my life. But I can’t really say that out loud . . .

To make things worse, Enzo loses his fragile temper with Mara because she can’t remember how to tie a bowline knot. He calls her stupid, and she begins to cry and runs off to close herself in her cabin for over an hour. Enzo stomps off toward his cabin, shaking his head.

The next morning our skipper is back to his amiable self, and Mara seems to have forgotten last night’s altercation. The two of them spend much of the day together on the aft deck, laughing and talking, with her patting him on the back like old friends. Even during her shift she remains distracted, flirting with Enzo from across the boat. Lisa seems visibly jealous, either for the loss of her friend or for not being with Enzo, or both. A boat is too small a place not to notice these things . . .

Despite the continued absence of dolphins, my training talks are still a daily routine. Trying to keep my crew's interest alive, I teach them about seabirds we've encountered and talk a little about oceanography.

Enzo seems to have made a habit of interrupting my talks with something he can't find aboard. I think he does it because he enjoys watching me get irritated. But I am not the only one. Everyone seems tired of his boorish demeanor and *spaghetti alla carbonara*, except Mara. Matteo also found himself the target of Enzo's fury. He left a glass of red wine on deck, staining the teak with a purple circle. Enzo started screaming insults, and unassuming Matteo, turning red like a tomato, had no idea what to do or say.

"*Idiota*," Enzo was shouting, throwing things around the cockpit.

That was it for me. It was enough that paying volunteers hadn't seen dolphins for almost two weeks; they didn't need to deal with a surly captain. I told Enzo he was out of line and had a responsibility to remain professional. Matteo's action was, after all, an accident. Enzo continued spouting off, shaking his head and glaring at me. I don't think he heard anything I said, but I felt responsible for looking after my crew.

The next afternoon we moor in a small, isolated harbor for the night. With an hour of light left in the day, Matteo and I drop the small inflatable dinghy in the water and go for a ride. "What are you doing?" Matteo asks me.

"I am sick of carbonara," I tell him, "I have another idea . . ."

I am looking for *patella*, not the kneecap variety but *Patella caerulea*, limpets that live on these intertidal rocks. When I was young, my mom and I used to collect these little gastropods, with shells shaped like the hat of a Chinese rice farmer, on the rocky beaches of Sardinia. They are delicious in pasta. We find an area covered with limpets not far from the *De Bolina*. I show Matteo how to insert his knife between the rock and the shell to detach it from its substrate. In less than half an hour, we've harvested a bucketful.

Back on the boat, Nella helps me quietly clean the limpets, put the pasta to boil, and prep a pan with olive oil, parsley, and garlic. The timing is perfect, and Enzo emerges from his cabin just as dinner is served.

"Tonight, *pasta con patelle*," I say, placing the steaming pot on the table. Everybody claps except the skipper, who shoots an annoyed glance in my direction. He sits grudgingly and devours a full plate, then helps himself to another. He doesn't say anything about the deviation from culinary protocol, but he is smiling as he leaves the table.

We are beginning our last day at sea, and we haven't seen a dolphin despite our time spent searching. Sitting on deck, all of us scan the surface for movements, any movement, in a last attempt. In the previous two days, there were no animal sightings at all. No fish, no turtles, no jellyfish—even the birds seem to be staying away. The water has been still, like a mirror with no apparent life in it.

The magical beauty of the Greek islands has not changed the fact that we've not seen a single dolphin. After all, dolphins are what we came for. As we pull into the familiar port of Otranto, I feel somehow responsible for not finding marine mammals. Rationally, I realize I have little to do with it, but I can't help feeling like I let the volunteers, and myself, down. Nella, Mara, and Lisa will stay for another fourteen-day turn, but Matteo is going home empty-handed, so to speak.

"I am so sorry we couldn't find dolphins," I say to Matteo as he waves goodbye from the dock with his backpack.

"Don't worry," he replies, flashing a broad smile I haven't seen before. "It sounds strange, I know, but I really had a fantastic time!"

4

Face-to-Face with Dolphins

"Ohiii!" Enzo cries out toward me leaning against the mast with his pipe, "I hope this is not another trip like the last one . . . I don't think I've ever had two in a row without dolphins. Maybe it's you, Maddalena, bringing bad luck."

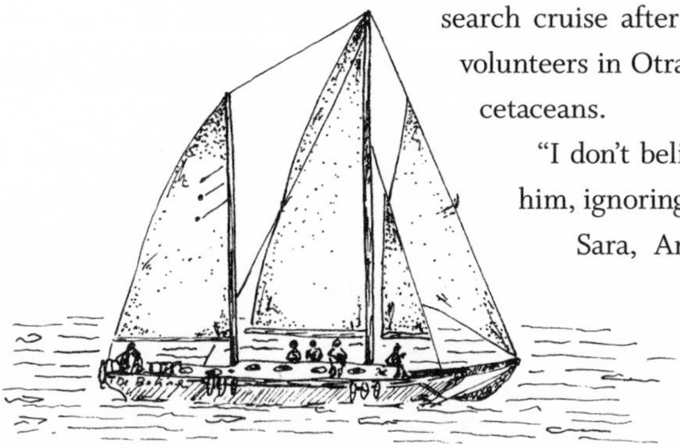

Four days ago, we had begun a new research cruise after picking up three new volunteers in Otranto. So far, no trace of cetaceans.

"I don't believe in bad luck," I tell him, ignoring his insult.

Sara, Anna, and Giuliana are new to our team, and they now join Mara, Lisa, and Nella on deck. They all work as secretaries for a large firm in Milan. Sara and Anna got their jobs when they were both eighteen, one to support her family, the other to escape her family, and now, at the respective ages of twenty-six and twenty-seven, they feel trapped in the tedium of their everyday routine. They both share the same dream of working with dolphins, and this cruise is their chance to finally see them firsthand. Giuliana is a couple of years younger and has come along with her friends for the ride. Overall, she is less interested in dolphins than her friends. They are all nice looking, in different ways, even though they share the ashen patina of lives lived under the tainted skies of Milan.

We stop at Lefkada for a day on terra firma. The island's gentle slopes descend to spectacular white beaches, and the air seems ever scented with a rustic combination of pine and sage. A large dish of freshly baked baklava gives off an inviting smell from a taverna in the port of Lefkas. We all sit, sipping coffee and eating baklava accompanied by a monotonous Greek tune. "We will be in Meganissi tomorrow, twelve nautical miles southeast of here. It has sea caves we can dive in," I say, trying to raise the spirit of my crew.

The next morning we are at sea again, searching for dorsal fins, a blow, any interruption in the still water's surface that, once again, seems devoid of any life. "Are we going to see them today?" Sara asks me, and all eyes turn to register my response. It's the middle of the afternoon on our fifth day and the team's enthusiasm is down. Perhaps I do bring bad luck . . .

Sitting down next to them, I say, "You need patience to study dolphins. There's never a guarantee of a sighting, much as we try to find them. Learning how to wait, sometimes in difficult conditions, is a skill any good cetologist needs." I have to admit, though, this time I am starting to wonder if we would ever find dolphins in these waters.

Then, suddenly, a scream: "Delfiniiiiiiiiiii!" and Nella is on her feet pointing to what look like ripples in the otherwise mirrorlike surface of the water. "Dolphins! At six o'clock," she repeats, still pointing. Enzo turns the boat and heads south in the direction of the dolphins. And there they are: about a dozen small black fins surfacing and disappearing underwater.

"They might be either common or striped dolphins," I cry out excitedly, straining into my binoculars to discern some characteristic markings. Well-timed, one dolphin comes to my rescue, jumping entirely out of the water. I can clearly see its coloration pattern; a gray-bluish dorsum, gray sides with a pronounced stripe as though painted with the flourish of an artist's brushstroke, a white belly, and a black band beginning at the eye and running down the flipper.

There is no doubt. "They are striped dolphins!" I say, relieved. The *De Bolina* is now a football field away from the dolphins, and Enzo waits for my directions on what to do. There are eleven individuals, all adults, judging from their size, and all heading south.

"Move parallel to them and not too close," I shout to Enzo from the bow.

My research team is ready to do what we've only talked about so far. I begin narrating what I observe while Nella logs the data onto the form I prepared weeks ago. It starts with our position.

"What was the latitude and longitude of sighting when we first observed them?" Nella queries. As soon as we saw the dolphins, Mara ran back to the cockpit and asked Enzo for our latitude and longitude from his SATNAV, writing them on a piece of paper.

"38°41'47" N, 20°45'32" E," he promptly replies.

Every fifteen minutes, we record the complete rundown of what we see. Nella reads me the rest of the questions:

"Date and time?"

"August 2, 16:30."

"Sea state?"

"Calm, no whitecaps."

"Species?"

"Striped dolphins."

"Group size?"

"Eleven individuals."

"Group formation?"

"Write 'tight,'" I reply. Lisa and Sara keep watch on the dolphins' direction.

"Dolphin direction?"

It's Lisa's turn this time. "They continue to move south in a straight line," she says.

"Yes, their direction is straight," I say. "Good job!" Nella writes it all down on the form.

As we parallel the school, the dolphins approach the boat to ride our bow wave. A kind of magic is beginning, and Nella, Sara, and I scramble into the net suspended below *De Bolina*'s bowsprit to be closer to it. Here we hang, with our heads upside down, just an arm's length from the water, the dolphins so near to us that we can almost touch their fluid bodies. One animal surfaces, grazing our fingertips, and a shiver runs down my spine. We are not speaking anymore, and it's clear all three of us are feeling the same intensity. From our perch, we can clearly see the individual detail of their painted bodies as they move effortlessly and gracefully through their salt blue world. For the first time, I can see the shape of their eyes under what seems their heavy black makeup. There is something there that I have not seen in my work with other animals, a kind of knowing or consciousness. It feels like they are as aware of us as we are of them. Time seems stopped.

Chuff. A striped dolphin blows a noisy mixture of air and mucus, awakening me from my surreal trance. It takes me a moment to realize where I am, that I'm the responsible biologist aboard, and supposed to be conducting dolphin research, not childishly fraternizing with the subjects. I reluctantly relinquish my space in the net to Giuliana and grab my camera to take some pictures as I resume the narration.

"These are *Stenella coeruleoalba*; it's their Latin name," I say excitedly. I tell the crew these are a pelagic species common to temperate and tropical waters all over the world. "They are not really known to ride bow waves, but I guess no one told that to these animals," I think out loud. "Maybe dolphins from different areas behave differently." I tell Nella to make sure she writes bow riding down on the back of the form as a part of what biologists call ad libitum observations or general field notes.

The dolphins continue their "unsanctioned" activity of riding our bow wave, as if they know that they are the primary protagonists of this day at sea, and that we are just their transient audience. One dolphin rolls sideways for a last glance upward, then peels off and downward, disappearing into the blue depths of the sea. Like a fighter squadron in formation, the other individuals bank to follow, and suddenly we are alone. Nella notes the time of their departure while the rest of us still search for a sign of their presence: another irregularity in the now-calm surface of the sea.

Between collecting data and our fascination with their presence, the two hours we spent with them passed in a flash. Now we are not ready to give up this wondrous first encounter, intoxicated as we were by their antics against the backdrop of an open sea. The sun has almost touched the horizon when we finally decide to take off our binoculars. Following my directions, Enzo proceeds south, then north, then circles several times around the area where we first saw the dolphins, but there is no trace of them; their flukeprints are gone. Enzo turns the wheel of the *De Bolina* toward Meganissi and the harbor of Vathí.

It's dark outside by the time we reach Meganissi, literally the "big island," although it is less than eight square miles in size. Tonight the mood aboard is different from anything I've experienced so far on the *De Bolina*. At dinner, the carbonara tastes better than usual, and all we talk about are the dolphins.

The enthusiasm is so contagious even our skipper seems in unusually good spirits.

Back in my cabin, cradled by the silence and the gentle rolling of the waves, I think about the day. I realize how, as I watched the dolphins from the net with my head upside down, for a moment I let go of that facade of scientific objectivity that I struggle to maintain. For a few minutes, I forgot about the collection of data, and, like one of my volunteers, I was absorbed by the magic of the encounter. I was proud that my emotions prevailed, even if only for a brief interlude. I promised myself I would never lose that sense of naïve wonder for the natural world, even if some might think it out of place in scientific circles.

At sunrise, no one talks about visiting the sea caves as we had planned. Instead the talk centers on where the dolphins might be. As the *De Bolina* glides through the oily water of the bay, leaving the chirping of the cicadas and the sleepy village of Vathí behind, we regroup on deck with coffee, data forms, and binoculars, ready to hunt for fins with the renewed enthusiasm of our first day at sea.

Early in the afternoon, another school joins us for a ride. Now that I've seen striped dolphins up close, it's easy for me to identify this new species on the spot by the pronounced difference in their color patterns. These are short-beaked common dolphins. They have a black stripe running from the lower jaw toward the flipper and a distinctive crisscross of colors on the side of their bodies. After three weeks on the boat with no dolphins at all, only to have back-to-back sightings of two species in as many days, we are all thrilled beyond comprehension. Our enthusiasm builds as the dolphins begin riding the bow of the *De Bolina* just as their striped cousins did the preceding day.

"Can you repeat the group size?" Sara asks me. Today she is the data recorder.

"There are six, all adults . . . write 'adults' in the notes."

I record their activities in my personal notebook, but it's difficult to keep my eyes on the paper and write down all the details. Following dolphins is like being in a fast-forward movie; it's all movement, very different from my Yucatán studies of lethargic, slow-moving sea turtles. Waiting for a female turtle to dig her nest, one could easily fall asleep on the beach . . . if it weren't for the airplane-sized mosquitoes sucking your blood. But here, among dolphins, one individual is glancing, another diving, another chuffing, another leaping against a golden afternoon sun, leaving a rainbow of droplets behind, and all at the same time. One just pooped, and a slimy yellowish-brown streak dissipates in all directions. I write this down, trying to keep pace with them.

"Look," Lisa says from the bow, swinging with half her body off the net, "they are so curious about us!" At the risk of sounding anthropocentric, I can't dispute Lisa's comment, and their glances do, in fact, seem directed at us. But to say they are curious is to impose a familiar human trait on animals that we do not understand, and I deliberately restrain myself from writing "snooping" in my notes.

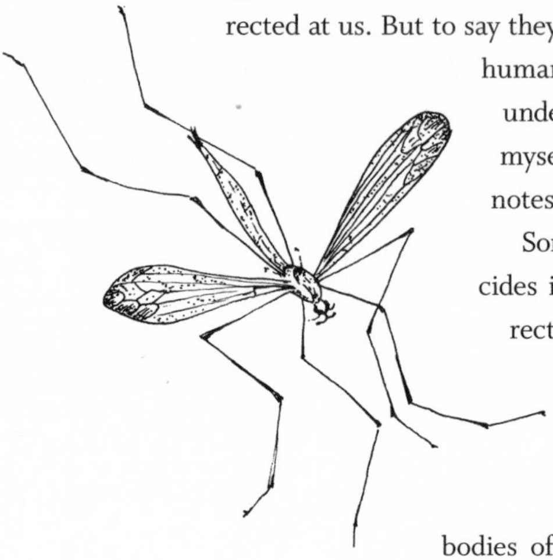

Some hours pass before the school decides it's time for a drastic change in direction, and once again, it's too late for us to chase them. As the *De Bolina* sails toward Ithaca, the sky turns orange-pink, and the water surface looks tie-dyed with the mauve bodies of purple-striped jellyfish creating a surreal oceanscape. The jellies move in a gentle rhythm through contraction of muscular rings at the bottom of their bells. These are *Pelagia noctiluca* (*pelagia* standing for "of the sea," and *nocti* and *luca* meaning "night" and "light," respectively). These marine organisms glow in the dark and, when disturbed, can become brilliantly luminescent. If handled, they leave a glowing mucous substance behind.

"How long do they live?" Sara asks me, taken by the grace of these wandering oceanic jellies. Sara is developing a sharp curiosity about nature and seems sincere in her desire to learn as much as she can about everything.

"Not too long, I think . . ." I reply, "and they often die in the rougher waters they encounter offshore."

It's late when we reach the island of Ithaca. The main town, Ithaki, was built in one of the world's largest natural harbors in the world, but by the time we get there, it's barely visible. Our long day of research has taken its toll on all of us, and after a quick dinner, we collapse, half dressed, in our bunks.

We leave for town shortly after waking up. We didn't make it this far during the first cruise, but now we're here, and Ithaca is nothing short of mythical. As we walk the ancient streets, it is as if we went back in time. One expects a tunic-clad Hippocrates or Plato to step out from behind every corner. But how could it be any different? This is the legendary island home of Ulysses. Everything breathes myth: the mountains, the scent of cypress and bougainvillea, the contrasting landscapes of green, smooth and harsh terrains, the caves of nymphs and the prolific ancient ruins. Even the drachma coins portray the image of the *Odyssey's* hero.

For the first time since we left Otranto, it's difficult to say good-bye to land. As the *De Bolina* leaves the emerald bay, I imagine Ulysses on the deck of his own vessel, departing this corner of paradise and sailing off toward the Trojan War. Though Ithaca was the essence of myth and magic, our next hours at sea will leave us speechless.

We are approached by another school of striped dolphins, but this time, they do not come to bow ride. For a while, the eight dolphins parallel our course maintaining a distance of thirty meters or so. Then suddenly, they launch themselves skyward in a prolonged series of synchronous leaps and bows without specific direction.

"Why do they all jump like that?" Nella asks me, without lowering her binoculars.

At that time, I had no idea why these dolphins were jumping. I knew spinner dolphins in the Pacific were thought to use leaping to rid themselves of parasites or, possibly, as a means of communication, but I didn't have a good explanation for striped dolphin jumps.

"They do it for fun!" Mara yells excitedly, taking shot after shot with her old Nikon camera. I find it hard to contradict her, given that these animals seem like a bunch of kids in their playground of the open ocean, but fun, like curiosity, is a sentiment that humans impose, and I somehow feel that any assumption on my part might cloud my ability to understand their behavior. But we are witnessing an acrobatic show, a circus of hops and inverted bows and long sequences of breaches. And whether it's true or not, we feel like it is put on for our amusement.

When the dolphins finally leave us, we sail north and, toward sunset, drop anchor in the protected hollow of a small cove. The water is so transparent one can count hundreds of vivid red starfish on the rocky seafloor. Enzo stands at the bowsprit and takes his shirt off.

"Oh no . . ." Jacopo says placing both of his hands over his eyes, "here he goes again!"

"What do you mean?" I ask.

"Watch and wait," says Jacopo.

Shirt off, pants off, underwear off, and there he is, buck naked, sporting an unflattering striped suntan. He turns deliberately around to face the girls, thinking, I suppose, to dazzle us with his manhood; he smiles and then, *splash*, dives into the water, reappearing at the surface a few moments later.

"Come on! Jump in, it's like a bathtub here!" Enzo shouts, waving his arms from a distance.

"We need to get our swimming suits," yells Sara.

"Don't be silly," Enzo yells back, "didn't you see . . . ?"

Nella giggles as she quickly peels off her top and slides into the water from the transom. A few seconds later, the rest of my crew are topless on deck, heading for the water. And there they are, splashing around together, our would-be wolf of the sea circled by half-naked volunteers on vacation. Jacopo and I remain aboard, not sure exactly what to make of our skipper's latest antics. We are all supposed to be the professional staff here . . . responsible, as it were, for the crew's safety. But it was hard to argue with Enzo's spontaneity and his unorthodox means of entertaining the troops.

"What do you think about a sea urchin appetizer tonight?" I ask Jacopo.

Armed with snorkels, gloves, and nets, we dive in to search for those black, spiny hedgehogs of the sea. Their ovaries are a delicacy, eaten straight from the shells with a drop of lemon. In less than an hour, we manage to round up a bucketful.

"*Aperitivi!*" I shout at my still-swimming crew, placing the bowl of sea urchins next to the open bottle of chilled white wine on the cockpit bench. One by one, dripping and giggling, they come aboard and, grabbing half a spiny urchin, suck out its contents.

The last days at sea had been memorable ones, made even more so through the complicity of the dolphins (and the wine). And tonight all boundaries have fallen away, leaving space for a warm sense of familiarity and comradeship among all of us, including Enzo.

This time, Lisa disappears with Enzo after the first round of appetizers. Again, eyebrows are raised, but Enzo's absence provides the perfect excuse for a well-deserved deviation from the usual carbonara fare, and Nella and I go below to prepare lasagna.

The next morning, we turn northward toward Lefkada. We sight a school of four bottlenose dolphins soon after leaving our anchorage. I recognize these animals at first sight, thanks to the encounters back in Yucatán with Superhero and the dolphins around our *panga*. Bottlenose dolphins are larger than common or striped dolphins, with stubby beaks and more pronounced "smiles." Their bodies are strong and vigorous with a uniform charcoal color that fades to almost a white underbelly. The dolphins swim beside the *De Bolina*, and, between bouts of taking data, I explain to my team how scientists identify individuals.

Back at home, I'd read a few scientific papers on dolphin identification by world-renowned cetologist Bernd Würsig and other authors, one of which explained how to identify and recognize individuals, using pictures of their dorsal fins. As each dorsal fin has its own distinct series of notches, generally received during social interactions with other dolphins, their fins work much like a fingerprint as a means of identifying individual animals. The technique consists of projecting the image of the dorsal fin on a screen, then tracing its shape on paper. Comparing fin tracings from each separate sighting with tracings from all other sightings, scientists can tell who's who among bottlenose dolphins. This system is more difficult to use on animals like common or striped dolphins because their fins are usually smaller. Bottlenose dolphins' dorsal fins are usually very different, and easy to distinguish.

I point to the dorsal fins of two individuals slowly traveling alongside to illustrate my argument. "Look at the large notch on the top of the fin of the individual closer to us, then look at the two notches on the fin of the second animal; see how different they are! It's like they have their names written on them."

"So are we going to photo-identify them? Are we giving them names? Can I name one?" Sara asks me, rapid-fire and ready with her camera.

"No," I tell her, "photo-ID is not part of our study. We need to have a better idea of what animals live in these waters before thinking about a photo-ID program." I explain that we are just taking pictures of every species we encounter during this research trip to confirm what we've observed. "Sorry, Sara, maybe we can name one after you in the future."

Sara looks disappointed, but not for long. A large individual has a notch on the bottom of its dorsal fin and massive scars along its body. It seems to lead the school as it positions itself in front of the others, slowly drawing them toward our bow. As we had already seen in our previous sighting, the dolphins turn sideways to take a look at us. Our eyes meet theirs. It may be all in my mind, but the more we stay with them, the more there seems some magical bond between us and our cetacean companions. The Greeks had a name for dolphins:

hieros ichthys, which means "sacred fish," and I am beginning to understand why. Greek and Roman mythology is full of tales that include dolphins in some way, often as the saviors of ships or men lost at sea.

Sara has decided arbitrarily to name the leading individual Boss. "How can they sleep and breathe at the same time?" she asks, distracting me from my thoughts.

"They are conscious breathers," I tell her. "They don't sleep like we do." To continue breathing, dolphins sleep with only half their brain, closing the opposing eye. By keeping one lobe awake, they are able to maintain normal functions like swimming and breathing.

In the meantime, lead by Boss, the bottlenose group leaves our bow and disappears. The sighting has ended. Several days pass without another dolphin sighting, but no one complains. We are in Corfu, on our way back home. For the first time, the sky is gray and gloomy, matching Enzo's mood on the morning of our departure for Otranto.

"Something wrong, *capitano*?" I ask. "Worried about the weather?" I am worried about his bad humor.

He nervously pours another cup of coffee from the galley's stove. "The forecast predicts a bora tomorrow," he replies, distracted. "Kind of unusual this time of year, but it's far away from where we are going."

Several years ago, with my parents, I was in Trieste on the northern shore of the Adriatic when the bora wind came through. The icy northerly gusts were blowing so hard I could barely manage to hold on to the railing along the waterfront promenade in the harbor of Porto Franco Vecchio. I still remember powerful waves crashing over the rocks with a roaring sound, and the salty spray stinging my face. I remember that when it finally passed, many of the smaller boats in the harbor had been blown ashore.

As we leave the fishing village of Kassiopi, at the northeast corner of Corfu, the sky is darkening and a chill breeze has picked up. We gather near the mainmast, wrapped in sweaters for a final training session. Enzo is below checking the forecast, and Jacopo steers the *De Bolina*, tacking occasionally for deeper waters. The wind is building, and with it come short choppy wind waves that begin to slap against our bow, throwing spray back over the deck and forcing us to retreat into the cockpit.

I go below to find Enzo furiously securing anything movable in the galley. "What's up?" I ask. "Looks like the conditions are getting worse."

"Not good . . . the report says small-craft advisory with a gale warning . . . winds up to thirty-six knots this afternoon . . . *Porca puttana*!" Enzo is visibly flustered and barking orders. "You can't trust these brainless weathermen! Tell your volunteers to stay out of my way and get their staterooms in order. Tell Jacopo to ease the mainsail . . . You go help him." As I reemerge on deck, I feel the wind picking up fast. I tell my team to go below and secure their gear. The *De Bolina* is sailing in seventeen knots of wind, heeling hard to port.

Five minutes later, our wind indicator reads twenty-four knots. There are whitecaps everywhere, and we are pounded continuously by waves. The girls are back on deck, nervously clustered around Jacopo, who is explaining that a sailboat can heel over without capsizing. "Don't worry," Jacopo assures them, "we can take a lot more than this . . ."

Enzo explodes from the companionway as if shot from a gun. "Out of my way means out of my way," he screams at no one in particular. "Nella, help Jacopo reef the main . . . Maddalena, you go take the jib down. *Porca puttana*!" My crew has shrunk into the corner of the cockpit except Anna, who is throwing up over the side. "Damn it, Anna," he continues his tirade, "go puke downwind!"

Enzo grabs the wheel away from Jacopo and starts the engine. Jacopo scrambles forward to reef the mainsail with Nella in tow. Sara guides Anna, now thoroughly green, toward the leeward side of the cockpit. The wind speed reads twenty-nine. Holding on the lifelines, I carefully work my way forward toward the bow. The deck is now completely wet and slick with salty foam, and the boat rolls as it pounds through the waves.

Seeing a panicky skipper on his own sailboat is not a pleasant sight, and it's making everyone nervous, including me. We have no way of knowing how dangerous our situation really is, and we all look to Enzo for reassurance, which he is not providing. My only option is to trust the captain and his mate and, in a worst-case scenario, follow the Mayday radio protocol to call for assistance.

Thirty knots of wind and getting dangerously close to gale conditions. I am precariously balanced on a wire holding the bowsprit in place, with one arm clamped around the sprit, pulling the sail down with the other as Jacopo, at the mast, lowers it to me. I am able to pull the sail halfway down before something snags and stops, so that I can't pull it any further. I can't secure it, and the jib is flogging wildly.

"Tell Enzo to hold his course; don't turn or I'll fall in the water!" I scream to Jacopo at the mast. He turns and yells something I can't make out, which Enzo seems to acknowledge. "Do you understand me?" I shout. Jacopo nods and continues lowering the sail.

I can feel the sail freeing up as I pull with all my strength. It is almost there when I look back to see Enzo making wild gestures in the cockpit. "Don't turn,"

I scream again. Without warning, Enzo turns and I am weightless; my body is flying in the wind, suspended over the water with one hand holding on to the forestay for dear life. Swinging like a monkey, I slam back into the boat and manage to grab the lifelines with my free hand just as I lose my grip on the forestay; I am catapulted hard onto the foredeck. My shoulder hurts, and I can see blood running down my hand onto the deck. I have a deep three-inch gash near my elbow. Jacopo runs to my aid, then secures the jib to the bowsprit while Enzo now motors the boat upwind.

Safe in the cockpit, I turn angrily to Enzo. "Are you totally insane?" I scream. "I told you not to turn." I would have hit him if my right arm hadn't hurt so badly.

"Sorry . . . didn't see you," Enzo quipped, looking elsewhere. Nella intervenes to take me below and bandage my bleeding arm. Fortunately, I do not need stitches. When I finally get back on deck, Enzo has gone below, and the wind and sea have calmed considerably. Anna is lying down in the cockpit, still seasick but recuperating. Jacopo is steering, and everyone else seems fine and engaged in a spirited discussion of this morning's events.

By early afternoon, the seas have laid down to a point that I can spend some time alone on the bow to rid myself of any leftover anger. I scan the horizon through my binoculars, looking for the westernmost of the Greek islands, Othonoí, where Calypso, the daughter of Atlas, held Ulysses captive for seven years. What I see, though, is not land but a fleeting glimpse on the horizon of an animal, probably some kind of whale. A minute passes and there it is again. It's a large animal, about six meters long, grayish with a cream-colored head, and a small, curved dorsal fin two-thirds of the way along the back.

"Whale dead ahead, one hundred meters!" I yell so everyone can hear me. My crew scrambles forward with their binoculars, but they can't find it. The whale probably made a deep dive. It is nothing I've ever seen before.

"What was it?" Nella asks.

"No idea, to be honest, I just got a glimpse of it," I reply. "Let's see if it surfaces again." Just when we are ready to give up, it surfaces again, then goes down for a shallow dive. I grab my camera, just as the animal surfaces for a second time and manage to take one photo, but it happens so quickly that I can't get a good look before the whale disappears forever.

Saying good-bye to these ecovolunteers in Otranto was difficult this time around. And so it was for the others who came aboard in my following cruises, with whom I shared the experience of observing dolphins up close. Although I did not know it yet, some of these people would find their way into my life as enduring friends.

When the season was over, I found it harder than expected to say good-bye to Jacopo. Our nightly talks under starry skies had become a regular part of my life in the last weeks, even if I hadn't noticed just how much I looked forward to these everyday encounters. I would miss him.

And looking back, I can admit that I would miss Enzo as well, even with a leftover scar on my elbow and the fact that I cannot, to this day, look at a plate of *spaghetti alla carbonara*. With his unpredictable mood swings and temper, to which I finally became accustomed, he was one of the most colorful characters I've met in my travels, and his tenuous friendship helped make my first experience at sea a memorable one.

Hardest of all, however, was leaving the dolphins. The more we learned about their lives at sea, the more some of us, especially me, found our lives forever changed. My battle with my insecurities was over for the moment, and I was going home feeling slightly more competent as a dolphin biologist. More than the rest, I would miss the silence, the golden sunrises and sunsets at sea, and the beauty, grace, and proximity of the dolphins.

At home in Padua, my phone rings. It's one of the research assistants from Tethys in Milan. "Hey, Maddalena, we got your film developed . . . Wow, you really saw a Cuvier's beaked whale! It's our first sighting ever!"

5

Adiós, El Palmar

"How about the Ría Lagartos biosphere reserve? There are at least two species of sea turtles . . ." Enrique says, taking another bite of his *camarones al mojo de ajo* and holding his one-year-old toddler across his shoulder like a sack of potatoes.

"I hate the idea of abandoning El Palmar," I tell him, "but you're right, we should check it out before I leave for Italy. By the way, has anyone seen bottlenose near shore at Ría Lagartos?"

Enrique, his wife, Monica, and I are having lunch together in Celestún to discuss the sea turtle research program at El Palmar. Enrique wears the open sandals and *guayabera** typical of the region, but they fail to hide his Germanic origin. He is of light complexion, tall, with clear blue eyes and ruddy blond hair. Enrique is the director of Biocenosis, an environmental organization based in Mérida, and my collaborator for the eco-volunteer program in Yucatán.

This is the end of another summer of the El Palmar project, and while I am generally pleased with the program, it has not

* Shirt commonly used by Yucatán men; also called the "Mexican wedding shirt."

been without its problems and challenges. Logistical issues, like the sporadic lack of provisions during the frequent tropical storms, were bothersome but surmountable. Other problems and hazards, however, that arose during the short life of the program proved more difficult to overcome. One such hazard was the scorpions.

Scorpions were at home at El Palmar. Under rocks, in shoes, in kitchen pots, or in the duffel bags of volunteers, El Palmar was their territory. Ordinarily, if undisturbed, they stayed away from us, but prudence dictated that volunteers learn, upon arrival, to shake out bags and clothes before using them. Sometimes volunteers learned the hard way . . .

In the first year, one of my crew, instead of shaking his jeans out as I taught him, grabbed his Levi's in a hurry and put them on. That's when something went wrong. In a second, his pants were off, and a scorpion fell out on the ground. We all watched helplessly while he ran screaming among the coconut trees in his underwear. By the time we managed to calm him down and get him in his hammock, he was pallid, feverish, and shivering. Two red welts from the scorpion's poisoned hook were visible on his leg, which was now beginning to swell, looking something like a giant sausage. His high fever lasted all day and all night. We tried painkillers and antihistamine, but nothing seemed to work. Just as I was getting ready to drive him to the hospital three hours away, he got better. Each year, at least one of us had a scorpion encounter. At times, in the midst of scorpions and coral snakes, El Palmar felt a little like a minefield.

The pristine beach of El Palmar was bordered inland by a seemingly endless expanse of coastal wetlands marked by stunning islands of mangroves and rain forest, called *petenes*, isolated by brackish mudflats or *blanquizales*. Enrique was studying the fish of these ecosystems. Many lived nowhere else in the world, and on a few occasions, my volunteers and I joined him to collect samples.

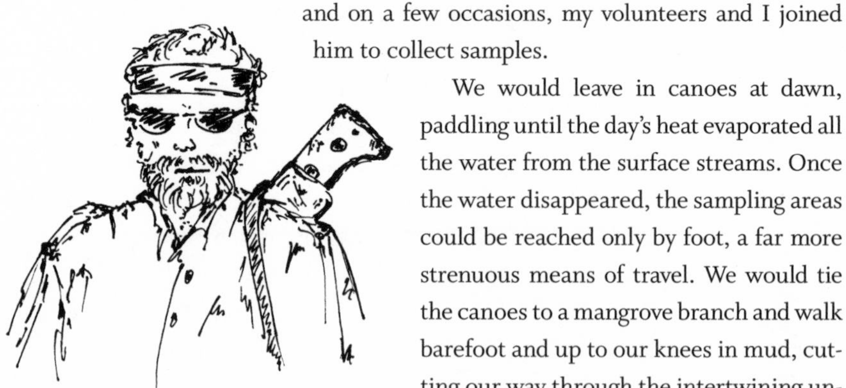

We would leave in canoes at dawn, paddling until the day's heat evaporated all the water from the surface streams. Once the water disappeared, the sampling areas could be reached only by foot, a far more strenuous means of travel. We would tie the canoes to a mangrove branch and walk barefoot and up to our knees in mud, cutting our way through the intertwining undergrowth with machetes. With Enrique and his broad-bladed machete at my side, I felt as if I had stepped into an Indiana Jones movie.

Here, in this fragile community of life, nature could be experienced at its best: herons, cardinals, motmots, hawks, arboreal snakes, crustaceans, and mollusks of all sorts were abundant. One could hear the screech of monkeys and catch a rare glimpse of an ocelot spying through the tangles of the mangroves' aerial roots. It was a pulsating microcosm, a biologist's paradise. But it was a demanding paradise to reach, and the perils of losing one's way could be severe . . .

One morning, I returned from my rounds to find a message on the kitchen table: "10:00 a.m.: Dear Maddalena, we went to the *petenes*. Back this afternoon. Ciao, L&N." Exploring the *petenes* with Enrique was already a challenge, but two volunteers with no previous experience, leaving late on a hot day for a place they could easily get lost or hurt, with no radio or machete, was nothing short of a potential disaster.

My research associate and I were preparing to go after them just as Enrique, along with two of his assistants, arrived from Celestún. Fully equipped, our search party set out for the *petenes* under a midday sun. For hours, we walked through thigh-deep mud, calling and shouting until we finally spotted the two of them lying exhausted in the shadow of a mangrove. They were utterly dehydrated, and their legs were cut and bleeding. When they saw us, they almost cried. Returning to our camp, I was struck by how easily things could turn dangerous in this remarkable place. I knew that living in El Palmar had its risks, but I thought the rewards outweighed them. Now I was beginning to think that I might be exposing my volunteers to far too many natural hazards.

And if the natural challenges at El Palmar were not enough, the appearance of drug traffickers sealed the project's fate.

Around 2:00 a.m., on an otherwise ordinary night patrol, a small group of volunteers and I had just finished collecting data on three hawksbill turtles when we noticed a lone female turtle wandering in circles in the lagoon, just past where we had been working. Evidently, she had walked over a sandbank and lost her way. Somehow the magnetic orientation that worked so well for her at sea seemed to have switched off on land, and we decided to help guide her back toward the sea.

Suddenly, at a distance, we saw the flash of lights approaching from the water. At first, we thought it might be poachers searching for turtles. It was no secret that hawksbills were illegally hunted along the Yucatán Peninsula, and anyone interested could buy artifacts made with their carapace in the markets of Mérida. We had never actually seen anyone hunting turtles at El Palmar, but we had found the trails left by the bodies of turtles after the poachers had dragged them from their nest sites. This isolated beach was an ideal spot to catch these valuable reptiles, as there were many turtles and there were no police at all.

We hid behind a dune to watch, uncertain of what to do next. The boat was much larger than a *panga*, and these people didn't look at all like poachers. About fifty yards ahead of us, on a dirt road running behind the sandbank, we noticed a large shrub-covered truck that we hadn't previously seen. Someone from the truck was signaling the boat with a flashlight, and a minute later two men carrying weapons and a few bags hopped off the boat as another four armed men met them halfway up the beach.

We clustered behind our dune that now seemed far too small to hide us all. We watched as the men exchanged the bags for something we couldn't see. After inspecting the bags, the four men headed back toward the truck, while the others disappeared into the sea like ghosts. Before leaving, one of the men in the truck aimed his flashlight toward the spot where we were hiding.

"He'll see us!" I thought to myself, ready to grab my team and run. But we stayed where we were until the man with the flashlight turned his attention back to the truck. He said something inaudible to the others, and then they left with their headlights off.

Later that morning, I drove to Celestún to report the event to the local police, but nobody seemed particularly surprised by my story. The police told me that drug traffickers used remote places like El Palmar for shipments, as they were difficult to patrol. The army had been attempting to curb these *narcotrafficantes* for some time, and the police promised to send military patrols at night, but we never saw them. They also told me to be very careful and to stay away from anyone we encountered on the beach, as the consequences could be deadly.

I could deal with scorpions and coral snakes, even work out the details of food shortages and severe conditions, but this was too much to subject volunteers to. Even though I wanted to stay, I had to face the reality that El Palmar had become a little risky for a project like mine.

"OK, Enrique," I concede. "We should definitely go have a look at Ría Lagartos. How about tomorrow?"

"*Bueno* . . . tomorrow is fine . . . and to answer your question, I have seen dolphins there."

As Enrique turns to order another beer, his child, still hanging from his shoulder, finds herself face-to-face with a bowl of freshly chopped jalapenos.

Her tiny hand moves so fast that none of us can stop her before she smears a handful of hot peppers all over her face. For a moment she is silent; then, turning ruby red, the toddler breaks out with an unremitting screech. She's on fire.

Enrique is panicked. "Maddalena . . . sorry . . . call you later . . . I've got to get her to the hospital . . . !" He collects his screaming daughter and lurches toward the door with his wife in tow. I pay the bill and head back toward El Palmar.

The next morning Enrique is back to his usual calm self, now that his daughter is sleeping safely at home. We are on our way to the Reserva de la Biosfera de Ría Lagartos, one of the prime natural areas in Mexico for diversity of habitats and species. Sitting next to Enrique, I stick my head out the window of his truck as he drives past a couple of vultures tearing at the leftovers of a dead cow. In front of us, a cluster of clouds stream east, like giant whipped-cream puffs in a bowl of azure sky. After passing the last haciendas and ranchos, the asphalt deteriorates into a road of sand, flanked by towering agaves. We cross an unstable creaking wood bridge where a man sips a *cerveza* while casting his fishing line into the murky water of the *ría*.*

As a white tail deer sprints away from us, the sky turns pink, and hundreds of flamingos take flight from the surrounding wetlands. Enrique says it won't be long before the white mountain of salt, adjacent to the town's saltworks, welcomes us to the fishing village of Las Coloradas, spectacularly sited in the heart of the reserve.

As we draw nearer, the radiant heat and dust make it look as though this little settlement sits suspended over a cloud. We pass through streets with no names, next to cabanas with doors wide open for ventilation, and simple roofs made from palm fronds. Inside, the furnishings are plain: hammocks, chairs, a wood table, and an old television—endlessly tuned to *telenovelas*.

There are no hotels, no real restaurants or stores at Las Coloradas, apart from a small bakery from which trails the scent of freshly baked bread. The *ganza*, a woman wider than she is tall, crosses the road in front of us with a tall stack of tortillas on her head. Her home is the closest thing to a restaurant, and fishermen returning from their day at sea often congregate there over a plate of fried *mero*† and a few beers.

We park the truck and walk past a dozen children playing soccer on a dirt field. Gangs of stray dogs are abundant, and one decides to follow us. A crowd of women with wide Mayan faces stares at us, as though we had just arrived

* A *ría* is a body of water with mangroves, flowing from sea into land and vice versa.

† Grouper.

from another planet. It's clear that tourists rarely frequent these parts. We smile and say hello, and walk toward the shoreline speckled by vibrantly painted *pangas* and the dead bodies of hundreds of horseshoe crabs, prehistoric-looking animals that migrate to these coastal waters during the breeding season.

The fishing fleet has just returned, and the fishermen cluster, drinking beer and chatting as they toss lobsters the size of cats, one after another, into plastic buckets. Fishing is what nearly every man here does now, but not long ago, the saltworks was what sustained the town. In Las Coloradas, most of the older men still bear the signs of those days when the sun, glinting off the salt crystals in the dry lake beds, devoured their skin, and carrying heavy sacks of salt permanently bent their backs.

We walk to the end of the old pier, a long wood structure that smells of guano and dead fish. A few double-crested cormorants and brown pelicans fight for a piling top to dry their wings, while the turbid water of the Gulf sparkles with tiny wavelets. I turn to look at a seashore that goes on as far as my eyes can see. It would be on this deserted stretch of sand that I would search for sea turtles in coming years. Enrique and I sit on the pier, and the salty sea spray tickles our legs.

"What do you think, Maddalena?" Enrique starts. "I talked to some people . . . we can get lodgings in town for at least twelve volunteers and food from the *ganza*." Without giving me time to answer, he continues, "You know . . . nature is unbelievable here too . . . jaguars, spider monkeys, and crocodiles just outside town, and in the lagoons you can see ibises, herons, spoonbills, hawks . . ."

I nod my head distractedly, but Enrique doesn't stop talking. "Oh . . . I asked around and they have hawksbills, green turtles, and, sometimes, even leatherbacks! Think about it: no narcos, no sleeping outside, and, every day, provisions at your doorstep. How's that for research-camp luxury?"

Enrique rambles on like a well-trained travel agent, and I think to myself how I will miss sleeping under the stars and being in such an isolated place as El Palmar. But moving the project to Ría Lagartos seems increasingly like the right choice.

A pelican plunges into the water and resurfaces with a large fish sticking halfway out of its beak. I haven't answered Enrique yet, and while we watch the pelican engulfing its prey, three bottlenose dolphins appear, moving along the beach in a tight formation.

"See . . ." Enrique says, boosting his pitch, "I told you there were dolphins!"

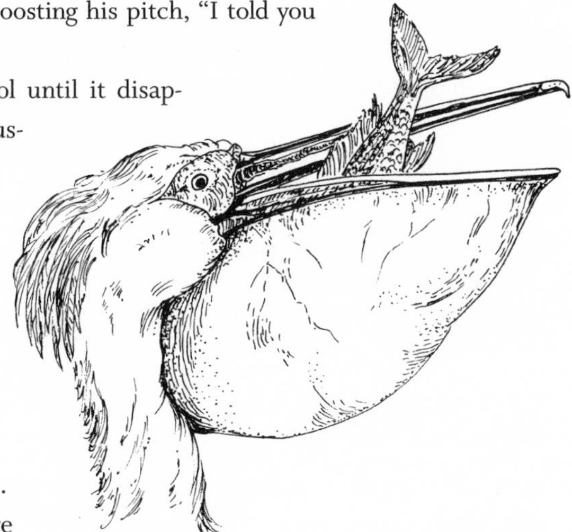

My eyes follow the school until it disappears at distance, then, excusing myself for a moment, I walk toward the fishermen still chatting near their boats.

"*Buenos días, discúlpame* . . ." I say in Spanish. "Do you see dolphins regularly here?"

"Oh, *sííí*, every day . . . just in front of the pier where your friend is," one man replies, while scratching his round, dark belly, ". . . or is he your husband?" Then, turning toward his friend sitting on the sand, he adds at full volume, "and they follow us when we go out fishing . . . right, Ignacio?"

Ignacio, who is obviously drunk, raises his head and, staring at me with crossed eyes, mumbles, "Huh? Ahhh . . . dolphins . . . yes, I whistle and they come to my boat . . . I am the dolphin catcher!" He laughs and collapses against the side of a *panga*. Our interview ends abruptly.

I walk back to Enrique on the pier. "OK, let's go back to Mérida and figure out the details of moving the project to Las Coloradas."

The Ría Lagartos reserve may not be El Palmar, but here was a chance to continue the nocturnal study of sea turtles and begin a daytime project on bottlenose dolphins as well. For a marine biologist wanting to spend time in the field, it sounded like the ideal combination.

After my first experience in the Ionian Sea, I had tried to start a dolphin study in El Palmar. I'd even drafted a research protocol for collecting data on cetaceans from shore. But the bottlenose there were not too cooperative, and the few schools I managed to spot from shore were barely discernible to the eye, moving too far away from the coast. In spite of my best intentions, my attempts at a land-based study had failed from the start. The situation at Las Coloradas seemed much more favorable, and I came away with rekindled enthusiasm.

On my return to Italy, my life takes a new direction when I decide to move out of my parents' home and in with my boyfriend. In the weeks leading up to the move, I have dreaded the prospect of announcing it to my parents, not only because of their traditional views about out-of-wedlock cohabitation, but also because the apartment we've rented is located in the hills outside Rome, some three hundred miles from Padua. Both Marco and I have been hired as journalists to work for a popular bicycle magazine. My boyfriend has taken a staff job, with the hope of working his way into being a career reporter. I, instead, decide to split my time between writing articles on assignment as a traveling photojournalist and continuing my studies as a marine biologist in Mexico and the Mediterranean Sea.

To my great surprise and amazement, my parents take the news remarkably well. Perhaps they have finally come to realize that the traditional path they expected me to follow is not the one I am likely to take.

That same winter I am recruited by Tethys as a research associate along the continental slope of the Ligurian Sea aboard their recently acquired sixty-five-foot motor sailer, *Gemini*. She is an ugly, round-bottomed ketch with accommodations for twelve. Seaworthy as she is, *Gemini* holds a reputation with all who have worked aboard as a sure source of seasickness.

This part of the Mediterranean is characterized by a high abundance of cetaceans, and Tethys scientists have undertaken both visual and acoustic surveys. Aboard *Gemini*, I meet Fabrizio once again. He is here to follow fin whales with a towed hydrophone array and to collect acoustic data on the vocalizations of these leviathans.

"Ciao, Giovanni's sister," he says, wired to his headphone and chewing on a sandwich as I peek down the companionway. Still chewing, he smirks, "Last time I saw you, you were green and lying in your bunk . . . I didn't expect to see you on a boat again, especially this one!"

"Well, I guess you were wrong," I reply, staring at the complicated panel of switches and lights in front of him.

"Come," he says in a conciliatory tone, "sit next to me, I'm not going to eat you," and he pats the bench with his free hand.

I move closer as Fabrizio explains the equipment I was staring at. He tells me how each species of whales and dolphins has unique vocalizations, and how fin whales produce short and loud pulses at a frequency of about twenty hertz, just below the human audible threshold. These pulses can be combined in various sequences of longer duration and repeated in bouts that can last for days. Using a hydrophone array towed behind our boat, made up of three hydrophones deployed at different distances in the water, Fabrizio can use these vocalizations to localize and track individual whales.

"You know," he explains, grabbing a cookie from a bag on the desk, "the sound emitted by the whale is received by each hydrophone at a different time because of the different distance of the instruments from the sound source. That's why we use an array."

He points out that it's this time difference, called time of arrival, that allows him to calculate correct distance and direction of the whale.

"Combining my acoustic way of tracking animals with other visual observations from deck gives us a great tool to estimate the relative abundance of whales here," he says. Then, excited like a child with a new toy, Fabrizio drops the cookie and points to a sound spectrogram.

"Want to hear a fin whale vocalization?" he inquires. "If I play back their sounds at a speed about ten times faster, it proportionally increases the pitch; otherwise you won't be able to hear it. The spectrogram is a visual representation of each sound that we use for comparison.

"The problem," Fabrizio tells me with insatiable enthusiasm, "is that humans hear sounds from about twenty to twenty thousand hertz—and these guys speak at one of the lowest frequencies in the animal world!

"Listen," he says, and pushes a button. There is a moment of silence, followed by an eerie, deep sound. It is like nothing I have ever heard before, almost unearthly.

"That's fantastic!" I say, listening carefully to the voice of the whale.

"You'll hear more of this whale talk during the cruise," Fabrizio explains, turning off the spectrogram and picking up the last chocolate cookie.

And it was, in fact, just as Fabrizio told me it would be. As we sailed, we tracked fin whales, using their "conversations" to follow them late into the night, long after our ability to see them had faded with the sunlight. We heard the clicks (short pulses of high-frequency sound) that different species of dolphins used to forage and the distinct whistling sounds they made when "speaking" to each other.

For the first time, I discovered that the world beneath that blue, infinite surface wasn't a realm of silence, as I had previously imagined. Instead, it was a world where sound travels five times faster than on land, where there was noise, and lots of it. There were sounds made by cetaceans to help them navigate and sense their surroundings, sounds for communication at a distance and to aid in locating prey. It was sound, for the most part, that cetaceans use to survive in the challenging, ever-changing, three-dimensional undersea world.

6

Conservation Awakening

I am sitting on the pier at Las Coloradas, waiting for dolphins and cradling my camera in my lap. I've spent most of the money I made last summer on a new 300 mm telephoto lens that will allow me to identify the animals I hope to find here.

The first team of volunteers will arrive in less than ten days for a two-week turn, and there is still much to organize. This morning, though, I am excited and ready to test my land-based methodology. I've revised my research protocol once again from the failed beach observations of El Palmar to fit the improved observation position of the Las Coloradas pier. I plan to collect data on group size, group formation, and dolphin activities from a fixed point. Some of the behavioral sampling techniques are based on the work of California scientist Susan Shane. It consists of observing a focal group of dolphins and recording, at three-minute intervals, their behavioral "states"—which are broad, defined patterns of activities—such as feeding or traveling or diving.

Taking pictures of dolphins as they pass close and parallel to the pier, I hope to identify each individual based on the notches on their dorsal fins. This was how I'd been able to recognize Superhero on my first trip to Mexico, by the pronounced notch in the middle of its fin.

"*Hola*, I'm Paulino," says someone in Spanish, just behind me. I turn to see a young boy, whose large almond-shaped eyes are overshadowed by his row of bright, oversized teeth. "What's your name?" he queries. "I saw you here last year . . . What are you doing?"

Before I can reply, he has taken a seat next to me. Paulino is fourteen, but his short Yucatecan stature makes him appear much younger. I tell him I am here to learn about the dolphins of Las Coloradas, but he is more interested in telling me about his seven brothers and sisters, all sent away to live with relatives because his father had neither the space nor money to keep them here. Paulino has only been to elementary school, and the farthest he's ever traveled in his life was Mérida, and there only for a day.

"So . . . are you married?" he asks, with a toothy grin.

"I have a boyfriend back in Italy," I explain. "We . . ."

"So . . . you are not married, then," he interrupts. "*Bueno*, that doesn't count then, because he's not here . . . so now you can marry me." The fact that we don't know each other, I'm almost twice his age, twice his size, live on the other side of the world, and speak another language doesn't seem to bother him at all.

"I don't think so, but that's very nice of you," I say, trying to be cordial, but he is not dissuaded.

"So, you think my offer is nice, then . . ." he says, beaming. "So, I will change your mind, seeing how you'll be here for months. Your boyfriend is nothing like me."

"No, that's for sure," I quip, just as a school of dolphins comes to my rescue. "I'm sorry, Paulino, but I have to work now."

The group is traveling east to west, unhurriedly moving toward the pier. I begin writing down my observations on the form I have devised for this purpose. When the bottlenose are almost in front of me, I take up my camera and peer through the viewfinder, attempting to snap pictures of their fins. One would think that with all the time I'd spent behind a camera, shooting pictures for my journalist job, I would have little trouble getting some good shots of the dolphins' fins. But my photos would later show that most of what I got was frames with nothing except water. This was much harder than what I had anticipated!

By the time the volunteers arrive at Las Coloradas, I have logged many hours observing dolphins from the pier, accompanied by the ever-present Paulino. He now calls himself my assistant, and as assistants go, he's remarkably adept at spotting fins from a distance.

Almost with clocklike precision, the dolphins visit the pier, leisurely traveling west in the morning and heading back eastward in the afternoon. And after a few days of practice, I become more skilled at anticipating their movements and keeping my Contax trained and prefocused on the spot where I expect them to surface. This produces better results than my first attempts.

Using a fixed platform like the pier, I was finding out, was not the best way to conduct a photo-ID study. First, the height of the pier didn't allow for the correct angle to avoid distortion of the fin. Second, there wasn't always enough time to photograph all the individuals in the school if the dolphins weren't cooperating: for instance, if they all dove before reaching the pier and surfaced too far away from me to get a decent shot. But even with this setback in my methods, I still felt I had the basis of a sound dolphin project in hand.

The Ría Lagartos reserve turned out to be a fairly good place to conduct research on dolphins, and over the course of several seasons, we logged over 240 hours sitting on that same pier, collecting data. It was the first dolphin project I'd ever organized from scratch, and the first land-based study on dolphin populations attempted on the Yucatán Peninsula.

The beaches of Las Coloradas also proved an excellent place for the sea turtle study. Unlike El Palmar, only frequented by hawksbill turtles, this area was visited by hawksbills, loggerheads, and occasionally leatherbacks. But while there were more turtles to study, there was more poaching and predation to deal with. I soon realized that more than research needed to be done here if these animals were to survive much longer. There was no one in Las Coloradas to blame for the lack of protection, as words like "threatened" or "endangered" had little meaning to the residents. Kids freely played war games with each other, using turtle eggs for ammunition, and the lights of the salt plant were so bright and glaring as to disorient turtle hatchlings at night.

But how would anyone in Las Coloradas know any better? No one had ever come here to explain the need or value of conservation. Turtles, to the locals, were an infinite resource of the sea, as they had been for as long as anyone could remember, and these villagers had no idea that the animals they took for granted might indeed be the last ones of their kind. Nor did they know that the delicate hatchlings inherited an instinct to crawl seaward from their nests

guided by the bright horizon above the sea, and that the presence of artificial lights could confuse their route, making them an easy dinner for terrestrial predators or causing death from exposure to the elements. To know these things, the people here needed education, and education was not the priority. Survival was.

Fearing for the sea turtles, I made an effort to include conservation and protection efforts along with the research studies at Ría Lagartos. The first step was to educate the locals about the importance of preserving these wild species. I needed to find ways of doing this that would be relevant to their lives, so they might find some benefit in changing their traditional ways. Not an easy proposition . . .

Some of my volunteers and I started by visiting the local school and talking about the importance of healthy ecosystems. We wanted to help the local children understand that their long-term livelihood depended on not taking more than the ecosystem could sustainably provide. Over time, I added a management component to the project, including an evaluation of the impacts on hatchlings from the salt extraction industry, followed by suggestions for alternative solutions to avoid further disturbance. And I began to "contract" some research services from the fishermen, hiring the use of their *pangas*, which provided alternative sources of revenue for the fleet.

For the first time, it had become clear to me that the pace of sea turtle exploitation was outpacing the benefits of pure research. I now believed that continuing in the traditional scientific vein of objective, detached investigation, without active conservation efforts, would only lead us to lose the very animals we had chosen to study.

By the end of the first season in Ría Lagartos, I had developed some basic curricula material, in Spanish, for use in the local school and partnered with local government fish-and-game management to protect turtle nest sites. But there was still lots of work to do, and I left Mexico with a renewed dedication to developing our conservation angle.

Returning home, I am confronted with an unexpected change. After almost ten years with me, my boyfriend insouciantly informs me that we are no longer a couple. He offers no explanation but expresses his wish to remain in the apartment and keep his staff job at the magazine. "You," he says, "have other options . . ." And with that, he leaves on an assignment.

I hardly know how to react. I am at first shocked and upset, but over the next several days, those feelings pass faster than I might have expected after such a long relationship, and I think less of him and more about what newfound possibilities my future might offer. Staying in Rome means pursuing a career as a journalist, with marine biology taking a backseat. But Marco was right about one thing: I had other options.

After a few phone calls, I am able to secure full-time employment between Tethys and Europe Conservation in Milan. My editor at the bicycle magazine also offers me freelance work that will help offset the low pay of my research job. Within a week of my return from Mexico, I find myself packing my belongings into my father's camper and heading to Milan.

I find a cheap one-room apartment in the heart of Chinatown and begin working as a researcher/accountant for the two organizations and a journalist for the bicycle magazine. After settling in, I also pick up some extra journalism work, writing occasional nature, science, and travel columns for national newspapers and magazines. All together, it's barely enough to pay the cost of this new life.

Waiting for the water to boil on a camp stove in my kitchenless apartment, I sit down to draft a new article proposal for a weekly paper. I am alone for the first time in my life, and I feel perfectly fine.

The next months are spent between Milan and cetacean surveys in the Adriatic, Ionian, and Tyrrhenian Seas. During this time, I get the opportunity to work with my brother, Gio, on a research trip to the French West Indies aboard *Gemini*. We log many days at sea, collecting data on species such as sperm whales, short-finned pilot whales, pantropical spotted dolphins, and an array of other cetaceans I'd never before seen.

A few times on that journey, my brother and I slipped away from our volunteer responsibilities to be soundless underwater spectators of a multitude of comb jellies. These creatures, transparent during the day, at night produce incredible multicolored hues that blend into a phenomenal light show of bioluminescence. In those moments, I felt like a tiny particle suspended in the water. I felt part of this dark and glowing liquid life.

The rich diversity of cetacean species frequenting the Caribbean tropical waters provides me opportunities to learn new things. At my brother's side, I can now distinguish one cetacean species from another, species that, until now, I met only inside pages of books. I learn how to tell one whale species from another at a distance, by observing the shape and size of their blows. In following animals from dawn until dusk, I see that one cannot rely on the coloration of an individual to identify a species. As the light changes over the course of a day, what may seem an ash-colored dorsum in the morning can acquire a dark blue hue at sunset.

In Milan, my days are filled with journalism and research. By now, it's become clear to me that there will never be much money working for Tethys studying dolphins. Not being born of wealth and wanting to remain independent, I'll need to continue doing other things to survive. The professional journalism career I'd begun in Rome would have surely offered more stability; in order to pursue it, however, I would have to give up what I am doing now.

But every time I go to sea, I feel that this is where I am meant to be. My fear of not knowing enough and failing as a scientist, my dread of seasickness, even my shyness, seem to dissolve whenever I take my place aboard a research cruise. At sea, there is a sense of peace one finds only in a few places on Earth. Some feel it climbing mountain peaks; others, while trekking into the vastness of a desert. But for me, it's the ocean. It doesn't matter if I'm sharing cramped, uncomfortable quarters with a dozen volunteers, or coping with the mood swings of an unpredictable captain. It's enough to sit on the bow, looking at the infinite cerulean surface, and everything seems in balance.

One day, my good friend and colleague Elena invites me to her house to meet a dear friend of hers from California. I have been looking for a way to cut my expenses and am considering finding a larger apartment to share with a roommate. Her friend is just moving to Milan from Tuscany, and Elena thinks it might be a good match.

Charlie is sitting on Elena's couch after eating the *penne al caviale** she just cooked for his birthday. He is tall, with broad shoulders and long blond hair tied in a ponytail. His blue-green eyes seem a bit shy, and he has a nice, kind smile. He speaks Italian well, and we talk for a while about the prospects of sharing an apartment.

Back in the States, Charlie was, among other things, a professional sailor, and Elena's family knows him through the sailing community. He's lived in Italy for a few years now and occasionally works with Enzo, delivering the *De Bolina* up and down the Italian coast. For one reason or another, Charlie and I decide not to be roommates. Instead, I remain in my miniature apartment, and Charlie rents a place across town, but our meeting is the beginning of a great friendship. In the months that followed, Charlie and I met frequently for dinners, concerts, or movies, sometimes staying up late into the night, walking the streets of Milan and talking endlessly about whatever came to mind.

Months later, Enzo is docked in Chioggia near Venice, ready to sail for Otranto. I know Charlie will be sailing down with him. Several boxes of T-shirts for the volunteers need to be delivered to the *De Bolina*, and I offer to take them. It's been days since I've seen Charlie, and I've missed our late night walks, which have become a regular occurrence back in Milan. As I get out of my car, I see Charlie working on the bowsprit of the *De Bolina*, and I realize that he is fast becoming much more than just a friend.

Time is short, and the *De Bolina* is behind schedule. Charlie and Enzo are busy making final preparations. Even so, Charlie and I manage a hasty dinner together, both struggling to overcome the shyness that keeps us from making what now seems a mutual intention clear. As I watch *De Bolina* pull away from the dock, I am swept with a mix of unexpected feelings: of sadness, excitement, and anticipation. Two weeks will pass before I'll see Charlie again in Milan for only a few days before I leave for the sea turtle research in Mexico.

As I drive from Venice to Padua, I feel as if a new world is opening before me. For the first time in my life, I am following my heart with little concern for practicality, but I am sure the path I am taking is the right one for me, even if it

* Penne with caviar.

may still be unclear. I know that I will somehow spend my existence in connection with the ocean and its creatures, be it as a scientist or protector, or some unknown combination of the two. I know also that my choices may not be the most sensible and may not conform to the traditional path of my upbringing, but I am surprisingly unconcerned. I am committed to following flukeprints wherever they might lead.

PART 2

Metropolitan Dolphins

7

California

Her name was *Scalawag*. She was powerful, and he liked her for how fast she sailed into the wind. Together, they were one at sea. Back in the City of Angels where he was born in California, Charlie found her again after returning from his stay in Europe. Without him, she got old and run-down, kept away from the ocean for so long. He carefully cleaned and caressed her back to life, until she became a racing boat once again. Then, untying the dock lines, the three of us left the harbor into the Pacific. Heeling to starboard, with the wind mightily filling mainsail and jib, *Scalawag* had come back to life.

I sit on deck with hands on lifeline, legs over the rail, and a new life at my fingertips. Behind me, an ocean apart and thousands of miles away, are my family, my friends, my country. I left them to be with Charlie, moving to an unfamiliar country, speaking a language I had never studied, meeting people I had never met. But it all came so naturally, it had to be the right choice.

We decided to move to the United States for several reasons. I wasn't earning much money in Milan, and the research job offered little future. On a visit to the States, I had made some inquiries within my field, and my impression was that the prospects in the United States, for my kind of work, showed more promise than those in Italy. Besides that, Charlie had gone through all his money after four years in my country and had a good job waiting for him back in the Hollywood film industry. Our initial plans were to go to California for a few years, check out the possibilities, save up some money, and see what the future would bring. For the present, California was my new home.

I still had supplemental income from my journalism connections. In fact, I was able to find even more work as a freelance travel and nature correspondent from the West Coast of America than I had back in Italy. One unexpected benefit of living in a foreign country . . . I also kept my sea turtle and dolphin project in Yucatán, agreeing to let the Italian office of Europe Conservation do the preliminary volunteer interviews while I organized everything else from California.

My parents, shocked at first, gradually accepted that this was my new life and that I probably would never make my nest near the family home, as so many of my peers had. The obvious fact that Charlie and I were happy together probably helped them accept my decision as well. I think, in their hearts, they had always suspected I wasn't cut out for the provincial life. Like an emigrant who's left the flavor of her land, taking reminders wherever possible, my baggage overflowed with parmigiano, pancetta, prosciutto, and some good Italian wine thrown in by my father.

Charlie passes me the tiller and fine-tunes the mainsail. A twenty-knot breeze makes *Scalawag* soar and crash through the waves of the Pacific. As our forty-footer heels with a fresh gust, a school of short-beaked common dolphins parallels our course. Soon there are so many dolphins it's hard to count them. They glide quickly at the surface, and I can barely distinguish their small fins from the spray streaking off the bow of our boat.

How different this sighting is from common dolphin sightings in the still waters of the Ionian Sea, where schools of just a few individuals meander slowly near the surface. Here it is all movement, as far as my eyes can see. For a moment, I have the impulse to grab my pen and paper from the bunk down below and take notes, but *Scalawag* disagrees, and her increased heeling keeps me rooted to the helm.

I struggle unsuccessfully to portray an air of nonchalance, as someone who's passed much time on sailboats and at sea, but I realize that *Scalawag* is another beast altogether from *De Bolina* and *Gemini*. Charlie, though, is completely relaxed as he resumes his place at the helm. He is drinking a Coke and tacks back toward land, casually pushing the tiller with his foot. With our change of direction, the dolphins gradually vanish into the sea.

"So," I ask, trying to shake the salt out of my hair, "do you see dolphins all the time around here?"

"Um, I'm not sure," he replies. "You're not really looking for dolphins when you're racing sailboats, you know . . . I've seen them bow riding a few times on yacht deliveries, though."

Sailing on *Scalawag* in the coastal and pelagic waters of the bay becomes our regular weekend date with the ocean. Charlie is in it for the thrill of the ride and to unwind from his intense week at work. As always, I love to be at sea, but after that first offshore sighting of common dolphins in the bay, these outings have ratcheted my interest up a notch.

Every weekend we see something. There are bottlenose dolphins traveling at a snail's pace near the harbor mouth, short-beaked common dolphins socializing offshore, Pacific white-sided dolphins porpoising off Malibu. In winter, we witness gray whales passing on their journey down to Baja from the icy waters of Alaska. There are sea lions resting on almost every buoy in the bay, and harbor seals balancing their chubby bodies on sunlit rocks just a stone's throw from the noisy traffic of the Pacific Coast Highway. Some are species I have never seen before now, and I am thrilled by the remarkable diversity of marine life I can find just a mile from my new home.

Over eighteen million people live in Los Angeles, but few know that the waters off this metropolis are frequented by a rich variety of cetacean species. To me, this comes as a surprise. After investing some time in preliminary research, I'm even more startled that it seems no one has ever studied dolphins or whales here. This is an enigma to me, but the more I think about it, the more I am captivated with the idea of being a research "pioneer" in these waters.

Santa Monica Bay is a perfect platform to conduct scientific studies, with its 180 square miles of semi-enclosed shallow shelf, stretching from Point Vicente in the south to Point Dume in the north. To the west lies the escarpment, a sharp, steep slope descending from the bay into open sea. Three submarine canyons cut deep gashes in the bay's seafloor, where the depth can plunge from the 164-foot average to over 1,500 feet in less than a nautical mile. A central plateau, called Short Bank, is the weekend destination of many sport fishermen, who gather in small powerboats full of fishing gear and six packs of beer.

My first "semi-official" observations are taken from the deck of *Scalawag*, but after several months of weekend-only outings, we decide to sell our sailboat and, with the proceeds, buy a modest secondhand powerboat for research. She is a twenty-two-foot Bayliner with inboard engine and a small cuddy to keep my camera dry. I find a slip in Marina del Rey, less than a hundred yards from our apartment.

I begin spending at least two days a week out on the water, searching out dolphins and whales and recording opportunistic observations on preliminary data sheets. Not having any background information on what species live here, my plan is to log time at sea, observing what's in these waters without any specific hypotheses in mind. Once I gain a better idea of the animals present, I'll decide what makes the most sense to investigate, and then design a research protocol accordingly.

It doesn't take me long to notice groups of bottlenose dolphins moving back and forth along shore on what seems an everyday basis. R. H. Defran, of the Cetacean Behavioral Laboratory at San Diego State University, and other researchers have already published work on the ecology of this species for the San Diego area, one hundred or so miles south of here. There are no data for Santa Monica Bay, but based on their results, it looks like the bottlenose dolphin population is highly mobile and might be sighted in my study area.

Some weekends, Charlie joins me for a powerboat ride. On a sunny Saturday, the two of us head to Catalina, a rocky island situated thirty miles south of our harbor. We've just passed the shipping lanes outside the bay when the wind picks up, driving a legion of whitecaps wildly toward our little boat. The Bayliner is more of a lake boat than a seaworthy vessel and takes hard punches from all directions, producing an unpleasant feeling, not unlike being cast into a washing machine. Charlie pilots the boat toward the small vacation port of

Avalon; I give up my search for cetaceans in the choppy conditions and instead focus on trying to keep myself stable aboard. A football field away, I catch the glimpse of a dark, slick profile.

"Did you see that?" I yell, compensating for the noise of the engine and the slamming of the waves.

"I saw something," he yells back. "What was it?"

"It's a whale, but I couldn't see what kind . . ." All I could make out in the midst of the swells was a small section of a dorsum; no head, no fin, no flukes.

We slow down and wait for the whale to resurface, but our boat is wallowing in the waves, and we're taking significant water over our bow. I clean my sunglasses, using my finger like a wiper, but it's a waste of time as we slam yet another set of swells. Then the swells subside just long enough for us to feel an earthquake-like rumble beneath us, as a massive blue whale surfaces next to our boat. The twin blowhole—the size of our cuddy cabin—rises out of the water as the whale exhales a powerful blast of misty air at least seven, eight meters high. For a moment we are speechless.

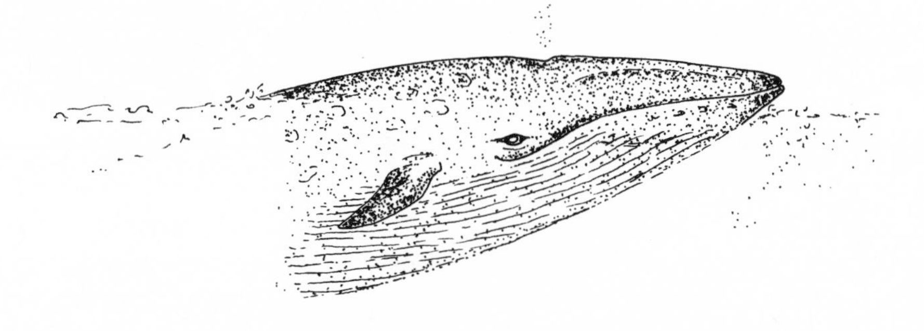

"Uh . . . he knows we're here, doesn't he?" Charlie manages, somewhat nervously.

"Probably," I say, fixated on the whale.

"Maybe we're too close . . . He could turn us into toothpicks without knowing it," says Charlie, still nervous.

"No, it's OK, just hold your course." I try to be reassuring.

We are in front of the largest animal on Earth, measuring over four times the size of our boat. I've never felt so small and frail in front of anything . . . but I've never seen a blue whale before. Neither has Charlie.

"Wow . . . amazing," whispers Charlie, practically stupefied, as the whale's fluke passes under our boat.

The long, bluish-gray body of the whale surfaces once more before disappearing. Thrilled and anxious, I search for the next blow. We are slammed again by waves so hard that I feel the familiar pang of nausea, but I am too concentrated on finding the whale again to deal with something silly like seasickness. We circle and wait, but the blue whale doesn't emerge again and, after half an hour or so, we head toward shelter in the port of Avalon.

The heart of a blue whale is as large as my old Fiat 500, and these creatures can weigh as much as twenty-five African elephants. As a child, I remember reading that a small boy could fit inside their largest blood vessel. But until I came face-to-face with one in the wild, I could never have imagined how impressive these animals really are. Even now, after many years spent in the field and other many encounters with the giant blue whales, this brief sighting remains one of the most amazing and memorable events ever.

After six months of collecting data and refining the questions I hope to answer, my research protocol is ready to go. I've modified the forms I developed in Greece and Yucatán, expanding them to accommodate the new data I want to collect. Now I have both a log and a sighting sheet. The log form covers information like weather and sea conditions, how animals are first detected (whether by naked eye or binoculars), and the first aspect one observes, such as a fin, a blow, surface splashes, or a concentration of birds. The idea is to create a picture of where we have been, so I can correlate that to where and when we sight animals.

The sighting form is for detailed data on the schools I see. Here I record information about cetacean activities every five minutes, to allow me enough time to observe the behavior of the dolphins and enter the data. I also use a more complete list of behavioral states than I did in Las Coloradas, and the new data include a list of behavioral events. These are single and recognizable movements or displays of an individual, like head up, spyhop, or bow.

Considering that sea lions and harbor seals are an integral component of the marine mammals living in the bay, I decide to include them also in my data collection. Sea lions are well investigated in their rookeries on land, but because of the added logistical challenge and accompanying expense of studying them in the water, less information is available on these animals at sea.

Once I've thoroughly tested both forms in the field, I realize that I can't possibly collect all this data by myself. It's time for me to seek some help—I must find and train a team of researchers, and with no money to pay them, they'll have to be volunteers. I will need a competent skipper who has knowledge of boats, a data recorder, and at least one skilled observer, other than myself.

A friend of a friend points me toward Travis, a young biologist with a stalwart passion for all things oceanic. Travis just finished his bachelor's degree in biology, and he's interested in working with me. Travis is already skilled in powerboats, having spent months on the ocean diving and collecting data for a small municipal aquarium. Moreover, he's a nice guy with a good dose of patience for learning what I need him to do.

At the time I first met Travis, my English skills were not too good, and my teaching abilities in this new language were not well honed, to say the least. In retrospect, I wonder what he must have thought on those frequent occasions when, out of frustration, I interjected Italian words, or words I just guessed at, making them up from a conglomeration of Spanish and German. But he never got upset and, somehow, managed to learn my protocols. I made some other inquiries through local schools and nonprofits, but finding volunteers was harder than I expected it would be. So, armed only with my new forms and my one research assistant, Travis, I decided to begin the research.

It's six o'clock in the morning. Wintertime. I am slowly starting to wake up as the cool breeze blows gently through our open window. I see a thick coat of fog covering the Los Angeles sky. Charlie shakes me a couple of times; he's leaving for work and heading toward the door. I kiss him good-bye, get up, get dressed, grab a quick tea and a croissant, and leave with my newly purchased Pelican case. The extent of my research equipment is a Nikon still camera, a 50–300 mm zoom lens, a dozen or so rolls of film, a pair of binoculars, a portable GPS (one of the first available on the market), and a brand-new handheld marine VHF radio. I'll use the GPS to record latitude and longitude at sea and to approximate the position of cetacean groups during our sightings.

As Travis and I leave the harbor, the exhaust smell of the boat blends with the intense stink of gull guano covering the breakwater like snow.

It's time to start the data collection. I pilot the boat past the breakwater as Travis looks over the log form. Given that there are only two of us aboard, at least for now, it's essential we both become proficient in all tasks aboard.

"So . . . what's the time of start survey today?" Travis asks me. "And . . . are we doing a coastal or offshore survey?"

"6:55 a.m. and it's an offshore," I reply, as the boat moves slowly through the water.

Today the water has an almost oily appearance, which reflects the clearing skies of the Los Angeles dawn. The sea seems devoid of all life. No dolphins, no sea lions, no birds, not even a surface ripple. Looking seaward to this immense desert of water, the contrast between city and sea is striking.

We cruise ten miles offshore and then turn back toward the coast . . . nothing. Reaching the south side of the bay, we spot a group of three dolphins. I write down the position where we first sighted the animals, then grab my data sheets and shift from log to sighting form, ready to take data. At the same time, I set my camera to photograph their dorsal fins and any interesting behavior they exhibit.

Another difference from my prior research is that the behavior of a dolphin school here is now recorded using a scan-sampling technique. Scan sampling involves taking an "instantaneous" sample of each individual's behavior before moving to the next animal in the group, repeating the same process until all individuals have been scanned. I try to conduct these scans as quickly as possible, from one side of the group to the other.

These three cetaceans are different. They are not like the other dolphins I've met before in these waters. These are large, robust animals with a bulbous head engraved in the middle by a vertical crease. They are light-colored, almost albino-looking Risso's dolphins. While juvenile Risso's can be difficult to distinguish from bottlenose because of their similar bluish-gray coloring, adult Risso's like these have a conspicuous skin discoloration due to the numerous white linear scars they acquire, in part, during social interactions with other individuals. By the time they reach maturity, their bodies have so many scars that older individuals look as though they are covered with a coat of white paint.

I had seen Risso's before in the Mediterranean, but the behavior of these California Risso's is different from those I met during my time on *Gemini*. In the Ligurian Sea, they were sociable, often turning on their sides for an inquisitive glance at us or riding our bow waves. Here all we can manage is a fleeting glimpse, not even enough time to get a decent photo. These animals seem shy and evasive, often diving as we approach, only to surface several hundred meters away.

Travis knows what to do when we sight cetaceans. First, he reduces boat speed to match the dolphins' movements. Then he parallels their course and records the velocity at which they are traveling. I have told him that sudden speed or directional changes should be avoided, in order to have minimum impact on the animals. Travis skillfully maneuvers the boat, approaching the animals as their dorsal fins disappear under the water. They resurface fifty meters ahead of us, and he speeds up to get closer. Once again, as soon as we are almost within photo distance, they plunge. We play hide-and-seek with them for almost twenty of these cycles, with us always carefully "chasing" the evanescent dolphins. We've been with them for over an hour, and I haven't yet taken a good picture . . .

I scan the horizon with binoculars, trying to guess where they might come up next. In the last fifteen minutes, the dolphins have been zigzagging like a bunch of drunks, not following any discernable direction. As they surface again in tight formation, this time only twenty or so meters away, there is a smaller, darker fin just a body length from the largest individual in the school.

"A young individual," I think out loud. But as we move closer, I see even more dolphins. We count ten new individuals and a calf, but these are not Risso's, they are bottlenose.

Together, they swim side by side like old friends, reunited. In the presence of their new companions, the Risso's seem less elusive than they were before. Slowly heading south, they swim at the surface, seemingly tired of their old hide-and-seek antics. Travis is now able to approach the newly formed, mixed school at close distance.

This interspecies mingling is something I didn't plan for on my sighting form. The form isn't set up for multiple-species data collection, and I scramble to jot my observations down on a notepad, keeping the activities of bottlenose and Risso's dolphins separate and straight in my head.

The bottlenose calf, which until now had been traveling next to what is likely its mother, moves away from her to take up an identical "calf position" near the largest Risso's. As if it were the most natural thing in the world, this acquired "parent" doesn't flinch, letting the baby bottlenose swim at its side. Travis and I watch this touching encounter as the calf gently rubs its head against the hefty right flipper of the Risso's, who slows down to ensure that its new "adoptee" can maintain its position.

Captivated by the observation of this new interspecies bond, I am reminded how, when observing animals in the wild, the most extraordinary things can happen when one least expects them. And once again during the course of my life in nature, I put my dry scientific objectivity on hold in favor of the empathy of the moment.

We continue to take data, still mesmerized by the sighting, until the two species break apart. An adult bottlenose with a falcate fin pushes itself between the calf and the large Risso's, and the bottlenose group makes an abrupt and simultaneous turn toward the shallow waters of the bay. We are once again alone with the Risso's. The afternoon breeze is starting to pick up, but this remarkable sighting can't just end. As we move on, following the Risso's, they resume playing their hide-and-seek game, and so do we.

In the weeks to follow, Travis and I would see mixed schools of Risso's and bottlenose with regularity. I'm increasingly enthralled with the idea of focusing my future studies on these aggregations. There are few investigations on mixed cetacean groups in the wild, and Risso's are not a well-studied species in general. If these mixed schools are a prevalent occurrence in Southern California, then this might be yet another fantastic research opportunity to fall at my doorstep.

8

Academia

The building is large, unpretentious, with razor wire over the gate, like any other warehouse here in the gang-infested neighborhood of Vernon, in South-Central Los Angeles.

After entering the empty parking lot, I walk around looking for some sign of the Natural History Museum Marine Mammal Laboratory. There is a freight container on the right side of the lot, and at the front of the building, there is a loading dock and a small door: all closed. I'm almost ready to leave, thinking I might have made a mistake with the address, when I see blood stains on the ground. Apprehensive but curious, I follow the red path toward the trailer-sized container, car keys in hand. The container is closed with a hefty padlock, and though I'm nervous, I can't resist peeking through the crack between the doors. As I push my face against the cold sheet of corrugated aluminum, a cold gust of death permeates my nostrils. I jump back and almost scream as a hand touches my shoulder.

"Maddalena, is that you?" a voice says. "I'm John, welcome to the whale warehouse!"

I turn to see an affable man sporting a bright Hawaiian shirt decked with surfers, waves, and palm trees. Trying to regain my composure, I shake his hand and introduce myself.

"Would you like to see what we hide in here?" he asks, seeing that I am still standing like a pole in front of the mysterious container.

"OK," I reply, not entirely sure that I mean it.

John unlocks the padlock and opens the door. As a cold, fishy odor makes its way out of the container, I see two semi-frozen carcasses of common dolphins resting on the floor. Rivulets of blood seep out of their stomachs, flowing out to the parking lot. Not the gang-related killings I was expecting . . . I feel like an idiot!

"Cool, no?" John says. "They stranded south of Palos Verdes a couple of days ago, and I went to pick them up last night."

John Heyning is the curator of marine mammals for the Natural History Museum of Los Angeles County and head of the museum's Marine Mammal Research Laboratory. He's an internationally recognized cetologist. For two decades, John has been in charge of recovering dead cetaceans along the California coast, often in gruesome condition.

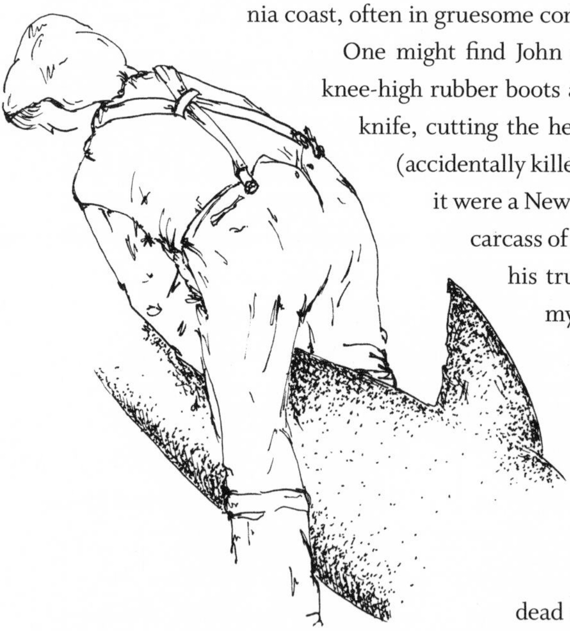

One might find John on Venice Beach, clad in knee-high rubber boots and armed with a flensing knife, cutting the head off a humpback whale (accidentally killed by a navy destroyer) as if it were a New York steak, or hoisting the carcass of a bottlenose on the back of his truck. Now that I'm back to my usual self, I see his large flatbed truck parked behind the container. On the back bumper is a sticker that says, "Save the Whales . . . Collect the Whole Set."

Working mostly with dead specimens, John has published scores of scientific papers and books on the evolutionary biology of whales and dolphins, including some milestone studies on the anatomy of the rare and mysterious beaked whales. John's idea was that, through dissecting corpses in the lab, one could get a glimpse of the lives of cetaceans that might never have been discovered just by observing these animals in the field. This, he thought, was especially true for evasive species like beaked whales. Even if I was never thrilled about working in a lab with dead whales, I had to admit John was right.

As John ushers me into the building, the sight is no less dreadful than the smell. I cover my nose with one hand, while my eyes take in the piles of jars stacked floor to ceiling, filled with organs, bone fragments, eggs, and undigested meals. I've never seen so many body parts and stomach contents in one place, and perhaps the only smell worse than this stench was from the poultry processing plants I passed in Delaware.

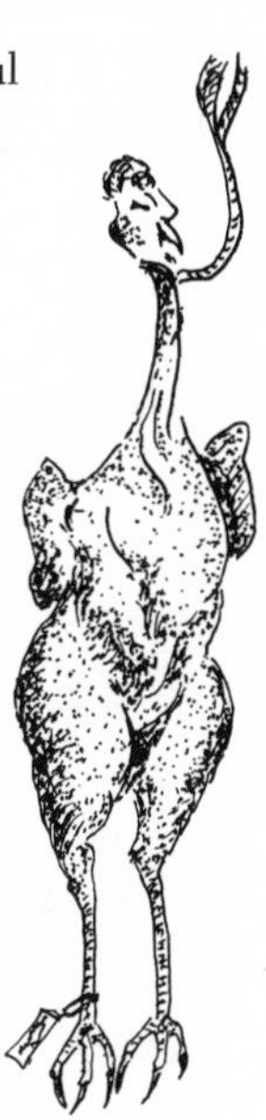

"You get used to it," John says, oblivious as he shuffles through a stack of papers on a desk. He picks up a stained coffee mug. "Would you like a cup?" he politely asks me, ready to pour another one.

"No, thanks," I say, half smiling, struggling to keep back my nausea. Coffee is the last thing I need . . .

I have no idea what to expect as John leads me through another door into the main warehouse. In front of my eyes lies a massive treasure of mammal skeletons, all perfectly preserved and assembled. There are animals of all types. It's amazing, something I have never seen before. In the corner are the twisted tusk of a narwhal, the six-meter-long skull of a blue whale, the keratinous baleen of a humpback, the large vertebra of a Risso's dolphin. There are skulls, antlers, and ribs belonging to a crowd of land mammals piled on large wooden and metal cabinets, and bones overflowing from plastic bags all over the stockroom.

John points out the bones of Bubbles, a female hippo that escaped from a safari park in Irvine, back in 1978. Bubbles evaded park rangers for more than two weeks before she was found hiding in a pond. The rangers shot her with tranquilizer darts, but Bubbles never woke up. Shortly after her capture, she died from the drugs, suffocated and pregnant. This was Bubbles's final resting place.

"Impressive, don't you think?" John says, sipping his coffee and looking at my expression. "We have something like twenty-five hundred whale and dolphin specimens here, without counting the six hundred or so pinnipeds. We are second, sizewise, only to the Smithsonian . . ."

I find it somewhat unbelievable that this difficult-to-find warehouse in the middle of Los Angeles gangland is one of the country's greatest natural-history resources and is entirely unknown to the general public.

Finally, John shows me his working area located near the loading dock. It resembles a scene from a horror movie, a sort of "torture chamber" with a large stainless-steel countertop, chains and hooks hanging from the roof, and blades and clamps of all sizes. Here is where the animals are dissected and sorted, and their flesh is industrially "steam-cleaned" from their bones. I am so captivated by all of it that I don't even notice the reek of the place anymore. I am getting used to it, just as John said I would.

"Do you think I can watch you work next time you find a stranded dolphin?" I ask.

"I don't see why not, " John replies, now busy cleaning some leftover blood off a knife. "I'll call you . . . and you can give me a ring the next time you're going out on your boat, if you have the space. It's always nice to get a breath of fresh air and see some living creatures for a change.

"Don't worry," he adds with a sarcastic smile, "I promise not to kill them!"

Leaning against a cabinet, upon which sits a dolphin skull grinning endlessly at me with its rows of identical cone-shaped teeth, I pass the afternoon with John, talking about research. I had contacted him a week ago to discuss my marine mammal project in the bay. After I explained what I was planning to do, he seemed interested in knowing more and invited me here to chat about the possibilities. John had done a lot of his studies on short-beaked and long-beaked common dolphins, both of which were present year-round in my study area.

John and one of his colleagues, William Perrin, analyzed over three hundred dead specimens of common dolphins from California, and discovered several differences, not only in their external morphology and anatomical characteristics, but also in their dietary requirements. Further genetic analysis showed that, although these animals live in the same waters and are consequently called sympatric, there was no gene flow between the two groups. The scientific conclusion that there were, in fact, two separate species with overlapping ranges was news. John, who had never actually spent much time observing these sympatric denizens in the field, was interested in what I was learning about their lives in the wild. I had a research boat and a field project ready to go in an area of his liking, and he had some extra time on hand. John thought this combination might provide some opportunities for all concerned.

"So, I hope you're planning to get a PhD with this project . . ." John says, taking me unawares.

Never, even for a moment, had I actually considered applying for a PhD! A doctorate seemed like such a big thing to me, especially coming from Italy. Over the preceding few months, I had made several trips to San Diego to meet with other scientists at the Scripps Institution of Oceanography. On one trip, I met with R. H. Defran at San Diego State University. R. H. offered to accept me into a master's program in his lab, focused on coastal bottlenose dolphins, which he had been studying for some time in the San Diego area, and I was strongly considering the idea.

A master's was a generous proposal and probably something I never would have achieved in my country just through an informal talk with a professor. Postgraduate degrees in Italy required that one be "connected," which is usually attained through working within the accepted academic or research community, and I was not a part of that group. At that time in Italy, one would never simply call up a professor whom one had not been referred to. American academia seemed more accessible, a place where one could move ahead based mostly on one's merits.

"Come on," John says, as though he had perfectly understood my hesitant expression, "you're way past master's material with what you know, not to mention your field experience . . ." He gives me a reassuring pat on the shoulder.

"Look, L.A. is not the best place to study dolphins because there are no marine mammal programs here," he continues, "but you should at least go see my friend Bill Hamner at the University of California, Los Angeles. Bill is a plankton guy and a world-renowned expert on jellyfish, but he has broad interests in marine science and oceanography. I'll call him and tell him he needs to see you."

The office of William Hamner is on the basement floor of the Organismic Biology, Ecology, and Evolution Department at UCLA. I am greeted by Peggy Hamner, who shares the windowless office with her husband and a few graduate students. She has a warm and welcoming face and shows me the way to Dr. Hamner's room. Everywhere there are piles of books, scientific magazines, scuba equipment, underwater camera cases, plankton nets, and desks overflowing with papers. Some of the equipment looks like it's left over from World War II, maybe even earlier . . . Dr. Hamner is in his early sixties and wears thick glasses and a white shirt under a velvet corduroy jacket that gives him the look of an old-school jungle explorer. Peggy introduces me.

"Come in, and sit down," he says, continuing to make corrections with a thick red pen on a printed page. "So, why are you here?"

Mustering the best English I can, I give him the short history of my education, my field experience, and my preliminary research in the bay. I summarize my discussion with John and ask if it might be possible to work here at UCLA.

After I finish talking, Hamner says, "Well . . . it all sounds good but, unfortunately, I can't take more graduate students right now." Holding up his hand with fingers extended, he adds, "I've already got five and I'm planning to retire soon . . . Really sorry, can't do it!"

I had spent the previous days preparing a résumé and a comprehensive research proposal for the meeting. Awkwardly, I hand them to Dr. Hamner, not so much because I think he wants them, but more because I don't know what else to do, as I have been holding them on my lap through the entire meeting. Our short encounter is over, and as I walk back across campus to my car, the aspirations for a PhD that John had sparked a week ago were evaporating fast.

At home, Charlie tries to cheer me up, telling me I won't have trouble finding another PhD opportunity if that is what I want to pursue. He suggests going out for a pizza, knowing that Italian food always helps pick me up when I am down. But I'm not in the mood. All I feel like doing is lying on the couch, listening to the sea lions bawling from the docks outside my apartment.

Around nine that evening, the phone rings. Charlie passes me the receiver.

"It's Hamner," he says.

Immediately I think I must have left something at his office.

"So, Maddalena . . . I've decided to take you for a PhD. You seem independent and your field experience is quite impressive."

I'm shocked and confused, but Dr. Hamner explains that after I walked out of his office, he looked over my research proposal and was positively impressed. He called John, who offered to act as an external adviser and evidently gave me a glowing report. This, apparently, was enough for Hamner, but with several caveats . . .

"Come to my office tomorrow at ten and we'll work out the details." He kindly says good night and hangs up.

The next morning in his office, Dr. Hamner has adopted a more familiar mode of conversation. He jokes about my strong Italian accent, and I can now see he has a sense of humor, albeit odd at times. He tells me how his interest in whales comes from the fact they eat krill, and krill is what he studies, among other things.

As our discussion gets more serious, we focus on the core content of my dissertation. We both agree on the need to assess some basic information about cetaceans in the bay, like occurrence, frequency, and distribution of species present here, before tackling anything more complex. I tell him about the interspecies associations of bottlenose and Risso's dolphins I've observed on multiple occasions. This, he agrees, will be one of the main topics of my dissertation.

"Now, to the conditions . . ." Hamner's tone takes on an ominous quality. "First, you need to take the Graduate Record Examination and an English proficiency test . . . and I expect near-perfect scores. Then, you'll need to become my best student, meaning your grades must be nothing less than excellent. You'll need to be focused and hardworking in your research, and you can't take forever to graduate. You should be out of here in four years or so. And . . . you will have to work as teaching assistant."

"But my English might not be good enough for teaching . . ." I interject, timidly.

"That's your problem." Dr. Hamner quips. "OK, meeting's over. I'll show you your office."

We walk across the hallway. "You'll share this with Myrna, another of my grad students," says Dr. Hamner, opening the door.

Inside, there are a couple of desks, both overrun with papers, biology books, dirty socks and underwear, stained coffee mugs, and open bags of chips. The bread crumbs on the dusty floor make a crunching noise as I walk to the window of my new office. Outside is a view of trash dumpsters. In the corner, there is a military cot, piled high with clothes, a sleeping bag, and pillow: all property of my new office-mate.

"It's nice," I lie.

"Myrna is kind of messy," Dr. Hamner says, rolling his eyes. "She sleeps here a few days a week . . .That's why there's a bed. You may have to fight for your own space . . . but, you'll figure it out. Now go get busy." I am dismissed.

As I leave the office, Dr. Hamner waves good-bye.

"Remember," he calls after me, "don't let me down, I made an exception in taking you . . ."

"How's that for pressure," I think to myself, walking up the stairs and out the door of my new department. I suppress a wave of my all-too-familiar insecurity. I'm both terrified and elated by this new and sudden turn of events, a turn that will become the cornerstone of the first long-term study ever undertaken in Los Angeles waters on the behavioral ecology of marine mammals.

Bill Hamner turned out to be a brilliant and erudite old-school academic with a mordant sense of humor that, at times, he used to test the limits of his students. He had a reputation for being kind of tough, but for me Dr. Hamner was always an outstanding adviser and a great help in supporting my efforts to find research grants and pay my school tuition. He later told me that his decision to take me as a grad student was influenced by his belief that I intended to conduct the research anyway, with or without the formality of the PhD. This belief is well-expressed in one of his recommendation letters for a fellowship I applied for, where Dr. Hamner wrote: "It is clear that Ms. Bearzi is going to study dolphin populations no matter what anyone else does, and we are attempting to provide her with an academic environment at UCLA that will enhance her investigations. Ms. Bearzi is the most focused, determined, and committed graduate student I have encountered in twenty-eight years of graduate education. She will do her research, even if she must fund it entirely herself. If the latter situation is necessary, the work will proceed rather slowly. If she can have some modest help, her work will be greatly enhanced." And so it was.

I decide on an official name for my research effort—the Los Angeles Dolphin Project, or LADP, a name at times confused with the Los Angeles Police Department, or LAPD. But either way, people seem to remember it.

Travis has since moved on to a new job as a marine biology teacher in New Jersey. I was sad to see him go. Over the months we had worked together, we had become friends as well as colleagues. To take his place in the field, I have found a fresh team of enthusiastic research assistants, the mainstays of which are Paul and Brigitte.

Paul has come to the project through his work with Bill and Peggy Hamner's educational outreach efforts. He's a teacher and aquarist for the UCLA aquarium, located beneath the Santa Monica Pier. He is tall, dark, and quiet, with a passion for the outdoors. But his truest calling is that of a teacher, and Paul is one of the finest and most talented educators I have ever met.

Brigitte is in her early twenties and is the first official LADP intern. She is the classic California beach girl with long blond hair and blue eyes. Brigitte works part-time as lifeguard for L.A. County Baywatch, while she's finishing her BS in marine biology at the University of Southern California. She is a terrific asset to the project, being attentive and capable, not easily fatigued, and gifted with remarkable physical strength. In her spare time, she plays water polo on her college team. Bill Hamner, who subsequently worked with Brigitte during one of his oceanographic ventures in Palau, was fond of saying, "When you need a man for the job, you call Brigitte!"

The hours we log following dolphins begin to number in the hundreds, as do the jokes and stories we tell each other to pass the time between our cetacean encounters. We are a well-matched team, and soon that same sense of complicity and comradeship I experienced in my days aboard *De Bolina* and *Gemini* returns.

Sometimes we pass entire days at sea without a sighting, returning home empty-handed. Other times the ocean is too rough or the fog is so thick that we can't even leave the harbor. But then there are times we have seven or eight dolphin or whale sightings in the course of a single survey: minke, humpback, and fin whales; orcas; Dall's porpoises; bottlenose, Pacific white-sided, and common dolphins. On those occasions, there is no chitchat and little space for words other than those necessary to conduct the research, each of us needing to accomplish our specific tasks. These are long days at sea, coming back to port only after the sun has slipped below the horizon, thoroughly exhausted and often cold and hungry. But these are the great days, the ones we will all remember best.

As we leave the breakwater, Paul calls the Baywatch lifeguards to let them know we will be working inshore this morning. We have an agreement that they will allow us inside the 300-yard boundary zone, as long as we inform them we are there and leave should they instruct us to do so.

In the silence of a cold sunrise, everyone takes up a position. Brigitte enters all preliminary data for the day's survey, while Paul scans the horizon for a fin or any unusual movement that might indicate the presence of dolphins. We are all quiet, and everyone seems rapt in thought as the boat glides north along the coast.

Paul is the first to spot a dark fin. A group of bottlenose dolphins are floating inshore, just outside the surf line. One individual surfaces slowly, taking a soundless breath. Three couples dive simultaneously. We edge closer, just enough to see them circling and ingeniously herding a small school of fish. A few animals float on the perimeter while the others feed. They are sentinels to warn against the appearance of other predators. They are deliberately patient as each waits its turn to feed, with no indication of haste or hostility.

One pair has returned to the surface. They are chuffing, taking noisy breaths, as though something is bothering them. Is it us? Paul checks the boat's position as they continue milling. I have briefed my team on the potential effect our boat may have on the animals we observe and explained how crucial it is that we respect how and where they live. Coming too close to where they are feeding can reduce dolphins' success in catching prey, and activities like bow riding are a good example of how dolphin behavior is modified, simply due to human presence. I've told my assistants how habituation to our presence may be reduced by carefully maneuvering the boat and leaving them their space. Learning how to do this comes only after spending significant time at sea and observing dolphin schools. Understanding how the animals move in the water is a critical factor. Paul has already become proficient at this task.

As our boat travels slightly away from the school, I see a dolphin winking at me, but it's probably only a wishful perception. It's just me, attempting to ascribe some human value to their actions. Sometimes I wonder how a dolphin might perceive the surrounding world. Dolphins can see fairly well out of water, but do they recognize us as individuals, or do they distinguish us in the same way a person with the wrong pair of glasses might, based on large and general features or overall shapes? Maybe, to a dolphin, I am just a good-sized, upright thing with some odd appendages . . .

With the dolphins moving slowly and foraging in one spot, we have a good window of opportunity to photo-identify the school. At the end of the five-minute period, Brigitte reads off the questions from the sighting form.

"Paul, can you tell Brigitte the group size?" I ask him, looking into the view-finder to get another shot of a dorsal fin.

"Nine individuals," Paul replies, "but I should count them again . . ."

He knows that the size of a group may fluctuate over the course of a sighting and that it's important to count the number of animals constantly to check for changes. This happens because bottlenose dolphins live in what are called fission-fusion societies. In this form of social organization, group size is not as fixed as it might be, say, in a wolf pack. Most dolphins are not solitary animals. They derive diverse benefits from association with a group, be it hunting or mating or raising newborns. But the members of any given group may leave at will, to form new groups or to join other existing groups. How long an animal stays with one group may vary from hours to months, even years.

A fission, or fracture, in a group may happen for a whole host of reasons. For instance, a few adult males may leave the group—still remaining in the same area—to hunt on their own, and then return to their companions a few hours later. To add another layer of complexity to this social organization, dolphin males typically have larger home ranges than do females and young. This is likely because males travel around looking for mating opportunities outside their own group. The result is an extremely fluid school, and the separation between neighboring schools is very difficult to recognize. As my new team of research assistants is beginning to learn, this type of fieldwork at sea can be complicated and demanding.

Sometimes a group might split because a dolphin becomes injured or sick. One or more individuals might leave the school structure to aid their companion in need. Scientists refer to this behavior as epimeletic. Healthy individuals rescuing distressed companions have been observed in the wild, and there are instances when this behavior has broken the species barrier . . .

We have just wrapped up our photo-ID work and have moved on to taking video of interactions and entering data on their behavior. The dolphins are still feeding in a circle near shore, when suddenly one individual changes direction, heading toward deeper water. It's followed, a minute later, by the rest of the school. We are so accustomed to following these coastal metropolitan dolphins back and forth, within a few hundred meters from the beach, that seeing them abruptly leave a foraging ground and change behavior surprises us.

"Paul, follow them . . . fast," I say, pointing in the new direction of the school. The dolphins have increased their speed and are still heading offshore. Paul pushes the throttle ahead, keeping pace with the dolphins, while Brigitte records this hasty change in behavior on the sighting form. We are three miles or so from shore when our group stops, forming a sort of ring around a dark object in the water.

"Someone in the water!" yells Brigitte, standing up and pointing to the seemingly lifeless body of a girl. For a moment, we are silent. Then, taking the helm, I slowly bring the boat closer. The girl is pallid and blond and appears to be fully clothed. As the boat nears, the girl feebly turns her head toward us, half raising her hand as a weak sign for help.

I cut the engine while Paul calls the lifeguards on the VHF. They tell us not to do anything until they get there. Our unanimous feeling is that if we don't act immediately, the girl will die. We decide to ignore the lifeguards' instructions, instead pulling the frail and hypothermic body on board. Brigitte directs the rescue efforts while Paul calls the lifeguards back and informs them that we have her aboard and are heading back to Marina del Rey as quickly as we can.

"She is cyanotic," says Brigitte, after a cursory examination, "probably severe hypothermia. We need to get her warm!" We manage to get some of her wet garments off and wrap her in a blanket. Brigitte and I take turns keeping her warm by huddling with her under the blanket.

The girl is around eighteen and probably foreign, because we can't seem to communicate. Brigitte tries in French and I, in Italian, Spanish, and German, but the girl is barely speaking, and none of us can understand what she's saying. I can't help noticing the plastic bag tied around her neck. It's sealed and seems to contain her passport and a folded handwritten note. Somewhere near the harbor, we meet up with the lifeguard rescue boat. We hand her off to them and head back to port.

A couple of hours later, we are all waiting outside the emergency room at the Marina del Rey hospital. The ER doctor comes out to talk with us. The girl, it seems, will pull through, and he thanks us for our quick action. He tells us the girl was vacationing in Los Angeles from Germany, and as the letter found in her plastic bag explained, she was committing suicide. If we hadn't found her, if the dolphins hadn't led us offshore when they did, to that specific place, she would have died.

Busy as we were trying to save the girl, we completely lost track of the dolphins. What might they have done with her if we hadn't followed them? Would they have tried to save her? There are many accounts of dolphins saving humans from death and disaster, either by guiding them to shore, fending off sharks, or helping them to remain afloat until help arrives. Many scientists think dolphins do not, in fact, save humans, because there is no hard scientific evidence to support these stories. But today I witnessed coastal bottlenose dolphins suddenly leave their feeding activities and head offshore. And in doing so, they led us to save a dying girl, some three miles offshore. Coincidence?

9

Cosmopolitan Dolphins

So far, most everything is going according to plan.

I have a massive amount of work to do between weekly surveys and attending lectures for my PhD. On top of it, I am trying to keep some money coming in by working as a freelance correspondent for several Italian magazines. I've settled back into academic life once again, and this time around, it is proving to be a rewarding experience. It is no longer the pure theoretical study that so frustrated me back in Padua. Now my studies at the university are directly tied to my research work in the field.

Since Dr. Hamner and I originally discussed incorporating associations of bottlenose and Risso's dolphins into my dissertation, the Risso's have . . . *poof*, evaporated from my study area. One possible explanation for their sudden disappearance was the occurrence of the strong 1997–98 El Niño, a poleward propagation of warm, nutrient-poor water along the coast of western America, caused by a breakdown of trade-wind circulation. El Niño occurs every three to seven years in the coastal eastern Pacific, usually resulting in a decrease of available food for some marine mammal species. This can bring about shifts

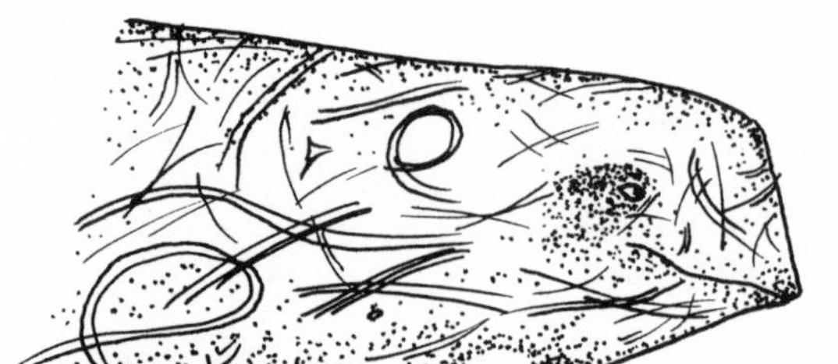

in the distribution of some populations, as they follow their prey to colder waters. Risso's dolphins are not good opportunistic feeders, and their diet consists mostly of squid; as the squid move offshore, so do the Risso's, leaving me with one less dissertation topic.

It didn't really matter. I knew one thing from my years of observing animals in the wild; I knew that if I was open-minded and patient, I would always find something enticing to explore. It's enough to open one's eyes to the natural world, and nature will offer endless possibilities for discovery. My possibilities were somewhere out there in the ocean, waiting to be found.

Charlie is aboard today. He follows a school of bottlenose into the surf line, making the lifeguard at Venice Beach nervous. In red shorts and with a throw around his shoulder, the lifeguard stands tensely by his yellow truck at the water's edge, holding binoculars in one hand and a bullhorn in the other, ready to snap into action.

Charlie and the lifeguard both gaze through their respective lenses, trying to anticipate each other's next move. The lifeguard is ready to use his bullhorn to send us away, and Charlie is set to call the lifeguard's superiors on the VHF to reiterate that we have their permission to approach dolphins inside the 300-yard boundary.

By now, most of the lifeguard staff know our LADP boat, thanks, in part, to Brigitte, who is one of them. But sometimes we come across an overzealous rookie that Brigitte doesn't know. Today we are fortunate, as the lifeguard gets back in his truck and heads off down the beach. We move on, negotiating the five-foot swell and heading north while Burbank, our new two-month-old Labrador puppy, settles into a nice, cozy spot on Brigitte's lap.

Burbank was Charlie's and my engagement present to each other. In the last few months, Charlie and I had been talking about getting married. Not that we thought a pair of wedding bands would change anything about our life together, but we both liked the idea of marriage. Our apartment building didn't allow dogs weighing over twenty-five pounds, so we had been searching around for a suitable puppy. One day, we went to look at some eight-week-old Labradors. As we played with the dogs, one little guy waddled over to Charlie, climbed into his lap, rolled belly up, and immediately fell asleep. My future husband turned to me with imploring eyes.

"I think this one's ours," he said.

Then, pausing for a moment, he lifted the puppy in the air as though he were weighing it, and added, "I don't think he's more than twenty-five pounds . . . Do you?"

And that was it. I drove home with our new dog on my lap. For me, this puppy was better than any diamond engagement ring I could ever imagine.

The school of bottlenose dolphins moves toward a dozen neoprene-clad surfers floating on their boards, waiting for a definitive wave to arrive. One dolphin approaches a surfer, then dives underneath his board. The surfer looks around for a sign of the animal, paddling with both hands toward where he thinks the dolphin will surface. But the dolphin emerges too far away to be followed by this slow-moving biped.

This is something I've witnessed countless times: Dolphins will approach surfers, swimmers, or kayakers. When they are almost close enough to touch, the animals typically disappear underwater, then surface a few meters away to resume their route, as if nothing ever happened. I have never observed any instance where physical interaction or contact has occurred. The dolphins may be curious, but in these waters, they do not seem to show any interest in actual contact.

We hold offshore at a safe distance until the school passes the last surfboard, then move closer to the animals to where I can photo-identify the individuals. They are sluggish this morning, even more than usual. One individual leaves the school, slowly approaching our boat. It swims with gawky movements. As it parallels the boat, it rolls upside down, and I can see that it's a male and that he's injured. There is an arm-length open wound on the side of his body, running from his right flipper down the flank. It's not one of the slight lacerations that dolphins can inflict on each other during normal physical interactions; even a shark couldn't have left such a clean slash as this. The cut seems human-made, likely from a boat propeller. From the bow, I click off several shots of both this individual's dorsal fin and his exposed wound. This is not the first time I have seen something like this.

As happened years before in my studies with sea turtles at Ría Lagartos, it was dawning on me that perhaps only doing research on cetaceans was not enough. There were too many critical environmental issues facing dolphins and other marine creatures all around me. I didn't feel like conducting my studies as though nothing peripheral was happening to these animals' ecosystems. I knew that I needed to do more . . . but I didn't know what the "more" was.

Charlie interrupts my thoughts. "Maddalena . . . where's Burbank?"

I scramble back from the bow into the cockpit, where Brigitte is frantically looking for our puppy.

"He was here . . ." Brigitte says, "just a second ago . . . on my seat . . . I was standing up collecting data . . . he was asleep . . ." She seems panicked.

Our Bayliner is not a big vessel, so it doesn't take long to search the whole boat. Nothing. Burbank is gone.

"Oh, my God . . ." says Brigitte. "He fell into the water!"

Charlie has turned the boat 180 degrees and is following our last course in search of Burbank. Motoring slowly, we all scan the surface for any sign of our puppy. We are all silent, trying not to believe that he might have drowned. How could we all have been so distracted not to notice Burbank's disappearance? No one heard a splash . . .

I am holding back my tears when Charlie abruptly stops the boat. "I heard something," he says.

There is a scraping sound coming from behind a bulkhead. I move my ear along the cockpit wall. There is definitely something there.

"I hear it," I whisper.

Charlie reaches shoulder-deep into the small space between the seats and the hull, and . . . there is Burbank, all rolled up and wedged snugly in a corner. We coax him out of his lair, and I hug him in my arms, as he looks up at me with surprised eyes and an obvious desire to go back to sleep.

"Thank God!" says Brigitte.

After that, we kept a close watch on Burbank, at least until he grew too large to disappear into the nooks and crannies of our boat.

Much has changed from the inception of my research. I feel I'm growing a little more confident as a cetologist, continuing to improve my methodology.

The data set for my doctorate is coming along nicely, I have a good team of assistants and a decent boat; however, as I log more days at sea, I see the increasing need to solve two different but related problems. First, I need to get more financial support for my work, which is becoming more time-consuming and complex from the standpoint of data collection. Second, I must find a way to do more than just research, as the more time I spend in the company of dolphins, the more I see a need for conservation.

After talking it over and deliberating for a few months, Charlie and I decide to found Ocean Conservation Society (OCS), a nonprofit organization with the broad mission "to conduct scientific research and educational projects leading to the protection and conservation of our oceans."

For me, it's a chance to continue my work as a marine biologist—with any luck, adding the structure for finding grant support and sponsorships. OCS, I hope, will allow me the flexibility to expand the scope of my research and to do something meaningful from an environmental perspective. For Charlie, who has always loved the ocean, this is a means to leave the affected movie business behind in favor of a different lifestyle spent out on the water in contact with nature. We had worked together once before in Italy, and now we hope that OCS will provide the vehicle for us to live together in the pursuit of something we both care deeply about. I think Charlie will make a good partner in the nonprofit. Where I understand the science, Charlie understands the business and administration.

Charlie's life was something out of a novel. He had done so many diverse things; a normal person might have needed a couple of lifetimes to accomplish the same. Charlie was a successful entrepreneur in several unrelated businesses, a blues musician and a music technician, a climber, a sailor, a biker, a cross-country skier, a political activist, a freelance journalist, a published writer, and photographer.

When I first met him, I had a hard time believing he was who he said he was. I thought, because neither of us had full command of the other's language, I might be misunderstanding the things he was telling me about himself. But as we came to know each other better, and especially after we moved from Italy to the States, I found that Charlie really was all of that, and then some. He was a restless type, always in quest of some new challenge or something new to learn. And he wasn't afraid to take risks.

I take the position of president at OCS, overseeing our research efforts, working on grant proposals and peer-reviewed publications, and maintaining the scientific permit that we need to study dolphins in the field. Charlie comes on as executive director, whose job is to run the business end of things, find money and board members, and assist with the research surveys. Charlie has a commercial captain's license, which helps us obtain appropriate insurance for our activities on the water and our assistants, something we desperately need, as the frequency of the outings is increasing exponentially. Together we begin working on environmental educational programs based on my dolphin studies. We hope to raise community awareness of the problems facing marine ecosystems and stimulate public action toward protection of the ocean.

Our decision is daunting, as it means that our collective income will essentially end as soon as Charlie quits his job and I stop working as a freelance journalist. But we have a little money saved up to cover our living expenses for a while, and we decide to move ahead.

"How difficult could it be," we wonder, "to get support for the research and the conservation of these charismatic species?" After all, everyone loves dolphins and wants to help them, and Los Angeles is an affluent metropolis with millions of residents. Our trepidation is soon overshadowed by our mutual excitement and enthusiasm.

One step at a time, we build a scientific advisory committee that includes experts from different disciplines, and we recruit board members, mostly from the business community. We also set out to find financial support for our work, writing letters and grant proposals and digging everywhere, from the private sector to large foundations, from federal funding to corporate sponsors. Suddenly my time on the water is now interrupted by time spent in front of a computer.

I walk Burbank along Venice Beach every week, often passing the site where the television show *Baywatch* is filmed. One day, while the crew is on break, the entire cast of *Baywatch* bombshells mobs me. Seeing me strolling with this adorable little puppy, they have all come over to pet him, led by Pamela Anderson with her signature red bikini and flowing crown of blond hair. If I were a man, I would have been in heaven.

Baywatch films in what is essentially my backyard, and the proximity of location, along with its worldwide popularity, gives me the idea to contact them to see if they might be inclined to support our dolphin project. After several calls and the mailing of many press kits, their publicist informs us that although David Hasselhoff is interested in our work, the show will be moving to Hawaii, so any official sponsorship is out of the question. I ask if any of the cast might be willing to come out on our boat, and the publicist says she'll be happy to extend the invitation.

"Hi, I'm Steve," says the voice on the phone. "Steve . . . from the cast of *Baywatch*."

His name doesn't ring a bell, but I'm not much of a *Baywatch* viewer, so I have no idea who Steve might be. But I know one thing: if Steve is in *Baywatch*, he's a celebrity, considering that this television series about Los Angeles lifeguards is watched by over a billion viewers a week. Steve wants to come along on the boat and check out my research work. He also wants to talk to me about getting involved, and I invite him on a survey for the following week.

As Steve steps aboard, we all notice his tanned muscular appearance and the *Baywatch* logos all over his clothing. He looks like a walking billboard for the show, but none of us recognize him.

At the north end of the bay, cruising close to shore, the water is so transparent I can clearly see long fronds of kelp sweeping softly under the surface like a giant heads of green hair tossed by the wind. Similar to a terrestrial forest, these dense kelp forests, anchored to the sea bottom, are among the most productive and dynamic ecosystems on Earth, offering a hiding place for predators, a home for an assemblage of organisms—from shrimp to rockfish to brittle stars and nudibranchs—and a substrate for an array of creatures to grow and reproduce.

We pass over the last canopy as a harbor seal surfaces among the kelp blades of this intertidal zone. The seal raises its silver and dark-spotted head above the water and periscopes the surroundings. Its head is round, with big eyes, long whiskers, and no earflaps. It stops for a moment, before retreating under the thick umbrella of algae.

"Write, one harbor seal . . . head up," I say to Mallory, a new volunteer in training, as she enters the data and asks Charlie for our GPS position.

Before turning south to head over the deep bathymetry of Point Dume Canyon, we pass another harbor seal, this time sunning itself on a pointy rock, its body balancing in perfect, nonsensical equilibrium.

A group of Pacific white-sided dolphins approach us from northwest, leaping at high speed. The elegance of these animals is striking, and this morning they are spectacular against the backdrop of the glassy water contrasted by a sky of deep blue. As a large male hitches a ride on our bow wave, he pushes aside a younger individual with quick strokes of body and fluke. I gaze through my viewfinder from my perch on the bow, and I feel as though I am swimming right next to him.

We collect data for over an hour, and when I am not busy dictating, I take time to explain the content and significance of our research to Steve. He's quiet and distracted, looking constantly at his cellular phone, which is dangerously close to losing reception.

"Careful, we're in the shipping lane," I say out loud to Mallory, who is now learning how to drive, coached by Brigitte. I watch as the speck on the horizon grows quickly into a good-sized oil tanker, moving fast on a collision route with our boat.

"*Sheeping* lane?" Charlie says. "Where are the sheep?"

"*Che simpatico* . . ." I reply, knowing he'll understand my meaning. "It's not nice to make fun of my accent." The volunteers don't think so; they are all giggling.

We traverse the *sheeping* lane and tag along with the Pacific white-sided dolphins. Steve doesn't seem to care about the dolphins and tells us he needs to get back into cellular range. He asks if we can return to port; it seems he has a critical afternoon meeting in Hollywood.

On the way back, there is little conversation aboard. No one is happy with our guest for interrupting such a great sighting. As we approach the breakwater, Steve comes alive.

"I just got a great idea!" he blurts out, flashing a smile with his dazzlingly white teeth.

He tells me that his character in *Baywatch*, which spanned only two episodes, was about to lose a leg in a freak boating accident, and shortly thereafter he would likely be cut from the show. Steve's plan was to pitch David Hasselhoff to let him stay on the show by becoming a handicapped dolphin researcher. He would work with dolphins from a boat, assisting in the rescue stories of each episode and bringing an environmental twist to *Baywatch*. We sadly realized that Steve was aboard neither to help the dolphins nor to support our work. He just wanted to stay alive in the cutthroat world of television. We never heard from Steve again, and his character was never resurrected.

A month later, a survey takes us forty miles offshore around Santa Barbara Island. After a final check of the morning's weather forecast, I hurry down to the dock and begin wiping the previous night's dew off my new LADP research boat. The old Bayliner had been showing its age in the form of recurring engine problems, and we have replaced it with a thirty-foot, twin-engine Pro-Line. It's a larger, more seaworthy vessel, with a cabin that offers protected space for our new equipment—which now includes two autofocus cameras, a video camera, compass binoculars, GPS, radar, and finally a laptop for data collection—a galley, and the luxury of an enclosed marine toilet. No more asking the crew to turn around while we awkwardly relieve ourselves over the transom, hanging on for dear life! What a welcome difference.

At 6:00 a.m. my crew wanders in, zombielike, with their respective cups of coffee. Bundled up in foul-weather gear, we leave the dock, hoping to get out to the island while the ocean is still flat. As we pass the breakwater, we are assaulted by the usual smell of concentrated bird guano that wafts over the boat. I think about the Captain Billy Tyne character played by George Clooney in *The Perfect Storm*, extolling the joys and beauties of leaving the harbor, as he inhales a deep and deliberate lungful of harbor air. My guess is that George must never have passed a breakwater like this one . . .

Laptop is on, binoculars and cameras are ready, and everyone seems alert. Joanna and Karyn are sitting on the bow, engrossed in conversation. Joanna is an energetic twenty-something hippie girl with pierced ears and tongue, and tousled, dreadlocky hair. She is new to the project, recruited by Brigitte. She is well spoken and shows a strong interest not only in our research, but also our public outreach work. Karyn is a few years older than Joanna and has been part of the team for about one year now. She, like Brigitte, is the consummate water girl and spends her spare time surfing and sailing.

I've just finished another ten-minute log entry, when I see a blow straight ahead of us. Two blows. Three.

"Whales! Twelve o'clock . . . two hundred meters!"

A trio of minke whales is circling a large school of herring. These are small whales, the second smallest of all mysticetes,* with males measuring around seven or eight meters in length. Their flippers have a characteristic transverse milk-colored band that reminds one of a waiter with a white napkin draped over his arm.

I set my camera to high speed as two of the whales lunge at the prey with their mouths wide open. One of them surfaces shortly after, raising its V-shaped

* Mysticetes (or baleen whales) are one of the two living suborders of the order Cetacea. The other suborder is the odontocetes (or toothed whales). Mysticetes have baleen plates for filtering food (mostly krill) from the water; odontocetes have teeth, and their diet is mainly composed of fish. The evolutionary origin of cetaceans is still a subject of controversy. Genetic analyses place them as close relatives of hippos, while morphological comparisons of fossils suggest that other "hoofed" mammals (like ungulates) are the progenitors of whales and dolphins. Whatever their origin may have been, cetaceans are the descendants of a land mammal that ventured to sea around fifty-five million years ago, becoming increasingly more adapted to the marine environment. In their evolutionary history, baleen whales and toothed whales diverged some thirty-five million years ago, giving us the array of cetacean diversity known today. Currently, there are eighty-six living species of whales, dolphins, and porpoises, but this number is subject to change, both for taxonomic reasons and due to species extinction (the baiji of the Yangtze River is now considered "potentially extinct" due to human impact on the environment).

head above the water, just next to our boat. We can clearly see their ventral grooves,* which have expanded enormously during the massive intake of water and prey, trapping the herring behind the sheets of baleen inside their mouths. As they close their jaws, the water is expelled through the baleen plates, leaving only the fish. While this whale-scale filter feeding is taking place, a dozen gulls circle over the whales in the hope of snatching some leftover herring, but these cetaceans have moved on. Fortunately, I have recorded the whole event on film.

I am almost ready to put down my camera, thinking the minke have ended their meal and disappeared underwater for their usual nine- or ten-minute dive, when the third individual explodes vertically upward from beneath the surface, closing its jaws in midair. It's a breathtaking sight, and all of us are frozen by the majestic display.

I've never had such a close view of a whale feeding. As I continue to photograph this spectacular behavioral exhibit, something is wrong . . . My camera doesn't stop at frame thirty-six, the shutter just keeps on clicking at high speed, and a wave of embarrassing doubt comes over me. Did I remember to put film in the camera? I look into the little window in the back of my Canon and . . . argghhh, no roll of film. I can't believe it! I didn't get a single image of the minkes, and this time the whales are gone for real.

Upset for not double-checking the camera, I sit silently on the bow as the boat glides over the building waves toward Santa Barbara Island. When we reach the island, we are welcomed by a large group of porpoising sea lions.

Santa Barbara is the smallest of the eight Channel Islands and looks more like a Scottish isle than what one would expect to find in California. A green carpet of grass covers the top of the islet, surrounded on all sides by steep rocky and sandy cliffs that descend to the sea. The water here is clear and blue, having lost all of its coastal murkiness.

This island is home to a large California sea lion rookery, and we can all hear their noisy doglike barks from a mile off. Many sea lions languish under the midday sun, piled atop one another. Through my binoculars, I can see a few large dominant males inspecting their harems. To the south, a couple of males are fighting with powerful lunges and ritualized movements. I watch as the larger one violently throws his entire body against his adversary. Both are covered with cuts and scars of this and other conflicts.

We drop anchor in a small sheltered bay to eat lunch aboard, surrounded

* The folds of skin that allow the underside of the mouth of a whale to swell like an expanding accordion.

by sea lions. Some thermoregulate or disappear for a short dive, while others curiously inspect our boat. Burbank lies on the bow with his nose toward the water. As he pushes himself dangerously close to the edge, a small sea lion lunges with teeth bared, snapping at the air just a few inches from Burbank's nose. Burbank pulls back quickly, then barks, as if to show who, indeed, is the alpha male here. For a while, they stare at each other, amusing themselves by barking and feigning lunges, but neither ventures from the safety of their own opposing medium.

As a stronger breeze is building and we decide to head back, Karyn spots a large school of bottlenose dolphins over a submarine slope. We draw nearer, and I notice that these animals seem larger and more vivacious than their coastal counterparts. They are making synchronous bows and spectacular leaps of the sort that one would see in an aquarium, coaxed by a trainer, soaring six, seven meters into the air. But here in the wild, it's something else entirely.

It's amazing how gracefully weightless these 250-plus-kilo animals seem as they effortlessly rocket free of their liquid world. These offshore bottlenose appear another species altogether from what I am accustomed to see inshore. They are all hops and bows and breaches. Physical contact seems continuous as they form acrobatic pairs or trios, then join together again to leap in a perfect row. To me, these animals are a picture of a freedom and independence perhaps unknown or lost to our modern human societies, needing nothing but open space and the fraternity of their kin.

As I see these dolphins soaring and plummeting in their three-dimensional medium, I reflect on how different their lives are from mine in the hectic world I live in. The first time I flew into Los Angeles, I was shocked by the city's spiderweb of freeways, stretching out as far as I could see in all directions. I'd never seen anything like it. Then, living in this metropolis with its mini and mega strip malls, its chaotic and interminable traffic, its endless restaurants and chain stores, my life changed. It wasn't like Italy, where I could just walk downstairs and buy fresh baguettes in the *panificio*. In Los Angeles, I need to get into my car, drive to one of the local football-field-sized supermarkets and choose among thirty different types of bread. Every time I go back to Italy, I am struck by how the roads and cities, which once appeared long and good-sized, now feel short and small. It's an odd feeling, as if the old world I grew up in has shrunk. In Los Angeles, I am lucky enough to live close to the water and spend time at sea, but as I watch the dolphins with no terra firma in sight, for a moment I envy the simplicity of their lives in the open ocean.

We come upon a group of exuberant juveniles playing among themselves. Three of them tow long streams of kelp on the end of their snouts. A few meters away, another couple of young animals break free every so often to prod each other's genital slits, and propel their playmate sideways through the water.

"Why are they doing that?" Joanna asks.

"Sexual play is an important part of a dolphin's life," I answer. "It's not real sex.

"They are learning something," I explain further. "Rubbing a rostrum into the genitals of a companion may seem weird to you, but it's the way they understand the reproductive state of another individual. What humans consider gross or inappropriate, like sex between an adult and a youngster, or incest, might be quite normal in the dolphin world."

We spent the entire afternoon following these lively "cosmopolitan" animals. We snooped into their private lives and documented details of their existence, all pieces of a yet unknown puzzle. This has been a magnificent day in the company of magnificent creatures.

This first long encounter with an exuberant school of offshore bottlenose now draws my curiosity deeper into pelagic waters. The eastern North Pacific seems to host two separate populations of bottlenose dolphins: a coastal, found less than one kilometer from shore, and an offshore, present in deep waters. Investigation of these less-known and more "cosmopolitan" dolphins will present some logistical problems, not the least of which is the amount of time I will need to spend in pelagic waters, but it also opens the door to a whole new set of unstudied questions.

Returning to port, we struggle against the waves that have now become substantial. The boat pounds through the chop, shooting streams of salty spray in every direction. Charlie stands at the helm, soaking wet and smiling, while the rest of us huddle in the cockpit, trying to stay dry and keep our lunches down. I guess this is the cost of our spectacular day at sea.

10

Party Time!

Today's survey starts early. We leave port as a spectacular November sunrise unfolds over Los Angeles. Charlie turns the boat offshore toward the outer reaches of the bay and deeper water. Flocks of western grebes floating on the ocean surface seem as though they might be attending a convention; brown pelicans soar in ordered formation. We are silent, scanning the surface for some sign, any sign: an odd movement of waves, a flurry of seabird activity, a blow, a dark fin emerging from the water. A sea lion pops its head up like a submarine periscope, looks at us, and quickly disappears.

We're now back in the blue desert of the ocean. Far off, we barely make out some wrinkling of the surface. We turn toward the disturbance in search of life. At first, we see what only appear to be whitecaps, but drawing nearer, we spot a few small fins. It's here, at the head of the submarine canyons, that we usually find animals. Ocean upwelling brings rich nutrients from the deep to the surface, and fish, seabirds, and dolphins often aggregate near these canyons to feed, sometimes numbering in the thousands. Dolphins' preference for bathymetric fea-

tures like these is not news. Scientists worldwide have observed several odontocete species congregating and foraging along seafloor reliefs, submarine canyons, and escarpments. These findings have shown that undersea topography, rather than water depth, is a key feature influencing the distribution of these animals.

We count two dozen short-beaked common dolphins in what seems a partying mood. Their quasi-continuous leaps and breaches splash the surface. A few ride our bow wave as others swim alongside, precisely matching our speed. In this festive atmosphere, we observe signs of sexual activity. There is much movement, and I can't ascertain whether these encounters are for mating purposes, social signaling, or simply amusement. Watching them and trying to put it in some human perspective, it looks like an orgy. A couple of pink dolphin penises get exposed, even flaunted in the dolphins' high acrobatic leaps. I try to keep my researchers concentrated on their tasks, but they find it hard to remain focused in the middle of this wild party. As they begin pointing and giggling, the distraction becomes contagious.

Suddenly, the dolphins dive and disappear all at once; the ocean goes flat, the festivity has ceased abruptly. My team is now silent, trying to make sense of this odd development and continuously scanning the surface in search of dolphins. A few minutes pass, and we find them again. But something is different; their "mood" has changed. They are determined now, swimming fast away from us and in rank formation. We feel deserted.

Something has happened. Something we couldn't hear or see, some underwater communication that swiftly changed their behavior. Listening to their vocalizations through our new hydrophones, we hear an intermittent myriad of random whistling sounds. Whatever the significance, the fact is that now we find ourselves behind them, trying to catch up. We follow the group for several miles, always maintaining our distance, videotaping, observing, and noting their route as they move along the length of the submarine canyon. These previously sociable animals are now powerful, resolute machines, which exhibit no curiosity or interest in us at all.

Ahead, a flock of brown pelicans, shearwaters, cormorants, and terns is circling, soaring, and plunging. Through our binoculars, we see more splashes as more birds arrive, swooping and diving in a frenzy. As we approach, the otherwise calm surface of the ocean comes increasingly alive with the wavelets created by diving birds and dolphin fins. There are now hundreds of dolphins, all seemingly ready for a different kind of party.

The frenzy increases. It's all flukes, fins, and flippers, as gray and white bodies streak through the water in all directions. It's hard to keep up with the changing directions as they dive, surface, then dive again. Running from one side of the boat to the other, trying to get it all down, we take photos and data, record videos and underwater sounds. As the dolphins' frenzy increases in magnitude, we too become more frenetic. To keep our energy level up, we grab pretzels and coffee on the fly.

"Leaping, ventral swimming . . ." I yell from the foredeck to Karyn in the cockpit, where she sits entering data on the laptop.

"Did you say leaping?" she asks, confused and a little frustrated by Shana, who is reading her the GPS coordinates of the school at the same time.

Shana nervously takes the helm, aware of the responsibility of maneuvering safely around dolphins. Shana is short, with dark, curly hair and pale skin. She's been with the LADP for several years now, taking time off now and again, to obtain a master's in marine biology and study at Woods Hole Oceanographic in Cape Cod. She has become one of our most trusted, passionate, and qualified researchers, and a good friend as well.

A male dolphin is bow riding. He gives me a glance, then leaves a stream of muddy-green feces. "Add poop in the notes," I say.

As strange as it might seem, data on dolphin defecation habits is valuable for cetologists trying to piece together these animals' behavior. Feces may even be used as a form of communication. It appears that when chemicals are released in the water, they inform other members of the school about that individual's level of arousal or reproductive state. So poop is part of our daily data-entry language as much as a bow, a leap, a spyhop, or a lunge.

Near us, just beneath the surface, a tasty fish "soup" awaits birds and dolphins alike. Cormorants dive deep into the water, and a few brown pelicans float nearby, gulping down beakfuls of fish.

Some of the dolphins push the large school toward the surface. The school seems a living cloud, swirling and reshaping itself continuously in spheres and tunnels as the fish struggle to find safety within their group.

We decide to deploy the new underwater video camera from the bow of the boat in an attempt to see what is going on under the waves. Wes is now helping Charlie set up the heavy camera equipment. Wes is thirty-something, tall, and with a large smile permanently fixed on his face. He comes out with us on occasion, usually when we need a strong, reliable hand with heavy gear or when we are doing something dangerous. Wes, like almost all my volunteers, is naturally drawn to the ocean. He is an avid kite surfer and takes great pride in showing off some new laceration or contusion from his latest adventure. Today, he has stitches in his leg and a nasty scrape on his forehead.

It's the first time we've used this camera offshore, even though Charlie and Wes have been working on it for months. Charlie wanted to build an underwater video system that could be controlled from the deck, thereby avoiding having to put a cameraperson in the water with the dolphins. I hope this will allow me to record their behavior during these feeding activities with a minimum of disturbance. It is a strange blend of technologies, but, oddly, it seems to work well.

Through the eye of the underwater camera, we are now seeing the activities that match our surface observations. We are heading into a building Santa Ana wind that is becoming stronger and stronger as the afternoon wears on. Wes is now firmly holding Charlie by the belt, keeping him in the boat, as Charlie attempts to follow the dolphin activity by rotating the pole.

"Oh, this is disgusting . . ." says Charlie. "I am going to puke with this stupid thing!"

"Not on me," Wes quips.

Charlie later explains that working with this camera is quite nauseating and disorienting, as what he sees in the viewfinder is the opposite of how he moves the camera. This unpleasant sensation, coupled with the movement of the ocean swells, is enough to make anyone seasick. Somehow Charlie manages to keep his lunch, but he and Wes dub the new video setup "the vomit machine."

I am busily taking pictures of dolphins and giving route directions to Shana, but for an instant, I can't resist the temptation of turning my camera on these two large men seemingly embracing in precarious equilibrium on the bow.

From the cockpit where I am working, I can see the video monitor that is hooked up to the underwater system. I can follow the dives of cormorants that I just watched slam into the water from above, catching fish undersea and breaking the school into two large swirling balls. They use their wings to propel themselves underwater as they do in air. It's a strange sight.

I continue to watch as three dolphins pass only inches from the camera eye. For a moment the image goes out of focus, but snaps back quickly, revealing something completely unexpected. A blue shark calmly swims in front of the video camera while dolphins and birds are still working over the fish school. Neither dolphins nor birds seem at all concerned by its presence.

Caged between water and sky by shark, dolphins, and birds, the fish have little hope. While glittery flakes of their scales float toward the surface, the fish break up into smaller and smaller groups, trying to flee. We watch as the last wave of remaining fish attempt a vain escape from a group of dolphins in pursuit. Some calves lag slightly behind their moms, still learning the methods that their place as top ocean predators will require. Like a ghost, the blue shark has vanished.

The burst of energetic activity has ended. The wind has subsided as well. Cormorants and pelicans have left the scene, except for a few birds floating on the surface, rolling with the gentle offshore swell. Sated, the dolphins move quietly off, perhaps slowed by their full bellies.

The ocean is calm and sparkles with silvery scales like a midnight sky filled with stars. Our net is out and ready for deployment from the stern of the boat. We use a plankton net to collect fish scales. It is a funnel-shaped, fine nylon mesh that is towed through the water, concentrating plankton (or, in our case, fish scales) in a detachable sample container at its end.

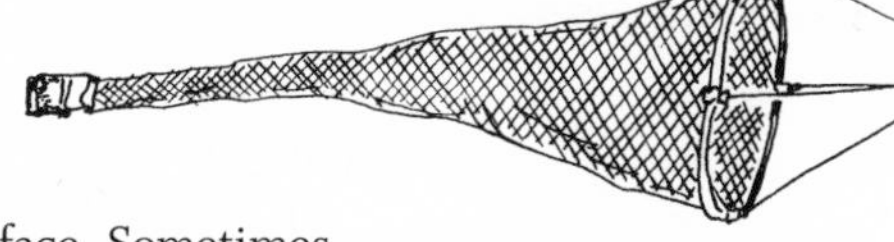

We slowly circle the feeding ground in search of scales, keeping the net suspended just below the surface. Sometimes fish are easily recognizable from the surface or from the underwater videos; other times, however, not being able to see underwater can limit our ability to determine the species. Scale identification under a microscope in the lab is another tool I use to get answers.

Shana is still piloting the boat; Charlie and Wes control the net depth and watch for the presence of scales. After ten minutes or so, we haul the net aboard and carefully detach the sample container, where plankton and other material have accumulated. I gently pick a few translucent flakes out with tweezers and drop them into a small plastic container filled with alcohol and labeled with survey date and sighting number.

I place the last container of scales in the boat's refrigerator and go back out to where Charlie is now washing the net with saltwater. While regular folks usually have a freezer jam-packed with meat and frozen vegetables, our home freezer is stocked with little transparent containers full of fish scales that, we hope, hold the secrets to the local dolphin diet.

I watch Charlie returning the net to its place in the side compartment. He seems so at ease now doing all sorts of research tasks. In the last few years, Charlie has learned a great deal about my work and about the environmental status of the oceans. This was another of his career diversions. I was astounded how fast he picked up everything about my field of study. For me, science was the natural state of affairs, having spent most of my time in classrooms and at sea. But for Charlie, this was an entirely new discipline, of which he knew little before meeting me.

I grew up with the conventional belief that academia is the place where one learns about science. Charlie, though, made me think that this isn't always the case. Einstein's words "The only source of knowledge is experience" fit my husband like a glove. Charlie was tangible proof that learning from experience and teaching oneself was possible, with only the interest, the curiosity, and, above all, the passion to do it.

Charlie takes over the helm and turns to cruise along the escarpment toward the other side of the bay, sixteen miles to the south. We spend a lot of time out here at the escarpment, where the relatively shallow plateau of the bay ends, falling off to much deeper water of the open ocean. We scan the horizon continuously but see nothing.

An hour goes by, then two, three . . . It is early evening, and still the ocean is silent and unmoving. We are getting cold and tired. Karyn's and Shana's eyes are bloodshot from the long hours spent looking at the reflective water or staring at a computer screen. We make small talk. Each of us has dropped into our own universe.

The sun is almost touching the horizon when a large black fin appears.

"Shark!" Shana yells.

We move closer, but the fin has descended into darkness. We see it emerge again after only a few minutes. It almost looks pliable, gently flapping side to side with the movement of the waves. A large pancake-shaped body becomes barely visible under the water. Clearly, not a shark. It's an ocean sunfish, or *Mola mola*, which, not disturbed by our presence, continues to bob up and down like a yo-yo in slow motion. We are all leaning over the side of the boat, captivated by this goofy creature. The sunfish is the heaviest bony fish in the world, with a puny brain the size of a walnut, a tail shaped like a rudder, and two long paddle fins at opposite sides of its laterally flat body. With a sluggish movement, the sunfish opens its small parrotlike mouth to grab a moon jellyfish passing nearby. Here, in the closing light of a spectacular sunset, this strange creature seems a lonely inhabitant of the sea.

In a quiet corner of the cockpit, I cinch up the collar of my foul-weather jacket to keep out the cold, moist air. Gazing out over the stern, I watch the phosphorescent trails of our wake curl outward, then dissipate into the dark surface of the ocean. I think back to earlier days and my preoccupation with whether I knew enough to be a good scientist or to lead volunteers in the field. Now, halfway through my PhD studies at UCLA, I still suffer from a nagging self-doubt, but I realize that I no longer worry as much about my competence in the field. Instead, I think about whether I'll do well on my doctorate defense, or how I will find the money to continue my work at sea.

I had been successful in training a great group of assistants, some of whom had become good friends. My research protocol was becoming more refined, and I was collecting data on a regular basis. I felt somehow proud of what I had accomplished, and more confident.

When I moved to the United States, I wondered how my life would change. I wondered what it would be like to live ten thousand miles away from my family and the friends I grew up with. I felt guilty in a way, for leaving them, and I missed them all. But in another way, I loved the excitement of shaping a new life.

Charlie and I had a unique relationship. It seemed as though we never tired of each other's company, even if we now worked together every day. There was an underlying solidity, strength, and mutual respect in what we felt for each other. We were lucky to have found each other, lucky to have become friends, and lucky to have become lovers.

As the lights of the city drew nearer and the smells and sounds of Los Angeles reached our boat, I realized that it didn't really matter where I was or what the future might hold. What mattered most was that I was following my heart and my dreams.

11

Desperately Seeking Dolphins

Dolphins may do amazing things or nothing at all. They may endlessly repeat the same action or disappear under the surface for a minute at a time. They may travel at high speed, socialize with gravity-defying leaps and synchronous jumps, or leisurely move back and forth just yards off the beach, "teasing" the surfers. Sometimes they ignore us; other times they ride our bow waves, playing gracefully or rolling sideways for a glance at our human forms, as if they too are studying us. But today, there is no trace of them.

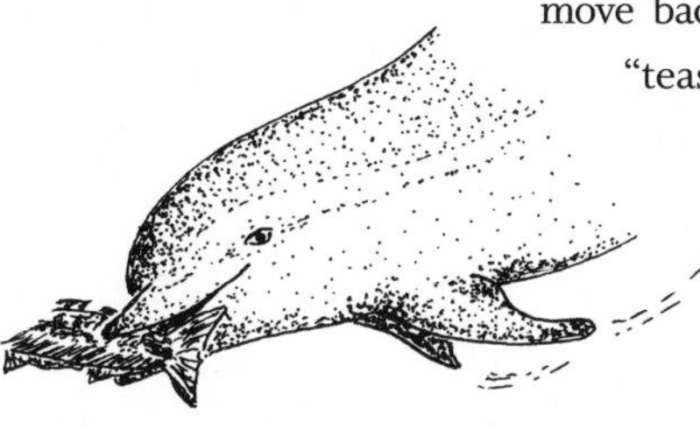

It's been six hours since we left our home port, first running parallel to the coastline in hazy conditions, then skirting the continental shelf in search of a fin. We turn offshore, running face-to-face into a dense wall of fog. Five pairs of eyes grow quickly tired from continuous scanning of the ocean surface in this reduced visibility. It's as if we are lifted on a veil of air, air so heavy one can cut it with a knife.

Our radar guides us toward deeper waters. The fog has a

soothing effect, canceling all noise and surrounding us like a blanket. Aboard there is silence, and we wait for that veil to lift, for that layer of impenetrability to bring us back to our familiar water and air. A ray of sun breaks the fog, and moments later a glorious day opens ahead of us. The water is flat like a pond.

I can see a silvery school of barracuda encircling what seems a shoal of anchovy. The prey have formed into a large, shimmering ball that expands and contracts in a futile effort to escape the powerful jaws and razor-sharp teeth of these adept predators. The aggressive hunters are only somewhat distracted by what zoologists call the confusion effect of their prey, and they dive into the mass of twisting fish, isolating smaller clusters of anchovies and consuming them.

Another hour passes at sea. Our coffee mugs are empty, and so are the two bags of pretzels we brought along, now dubbed the semi-official food of the LADP. Shana munches on her peanut-butter-and-jelly sandwich.

"One California sea lion head up," I say, looking at the chocolate-brown head that emerges in front of us.

"Two sea lions head up," Shana corrects me, pointing to another individual surfacing to starboard.

"Six more sea lions swimming," Charlie says, motioning toward a group moving steadily in our direction.

The six sea lions join the other two animals near our boat, and the whole pack swims off together, porpoising north in decisive fashion, as if they are late for some important ocean meeting. Usually we don't follow pinnipeds, but there are no dolphins around, and the pinnipeds are moving along our transect line, so we speed up to stay with our new group of torpedo-shaped friends.

Every few hundred meters, they pause and spyhop at the surface, as if they are searching for something; then they resume their porpoising. After ten minutes, we see a group of approximately thirty short-beaked common dolphins swimming steadily northeast in a tight group. The distance between the two species narrows as the handful of sea lions increases its speed and adjusts direction to match that of the dolphins. If the dolphins want to evade the sea lions, they could easily do so, being capable of speeds in excess of sixteen knots. But the cetacean school is not at full power and moves decisively along at an approachable velocity of about nine knots.

"Hmmm . . . purposeful, porpoising porpoises," quips Charlie. Everyone ignores him.

So here we are, chasing sea lions chasing dolphins, which are probably chasing something else. Suddenly, the dolphins fan out into a rank formation and increase their speed. The distance between them and the sea lions is now increasing, leaving the sea lions (and us) lagging farther behind.

Where did the dolphins go? For a moment the sea lions seem lost; they stop, sticking their heads out of the water for a better view. Then they adjust their route to follow the dolphin school, speeding up to close the gap.

"We should get this on video," I tell Charlie.

The dolphins slow down, closing ranks around a large school of fish. This must be what they were all searching for. With elegant nonchalance, the sea lions join the dolphin school, feeding together on the swirling ball of anchovies. A few dolphins lunge upside down after their prey, leaving a sparkling trail of scales where their target fish once was.

A sea lion moves next to a dolphin, almost matching the dolphin's more graceful movements. But the dolphin pivots effortlessly and dives into the swirling fish ball that now disperses in four different directions, leaving the less agile sea lion a step behind, and alone. The abandoned pinniped spyhops, striving to determine the latest position of the shifting foraging ground. Then, seemingly imitating the behavior of another dolphin next to it, it attempts a high, lateral leap with one flipper slicing the air as would a dolphin's dorsal fin.

From my surface perspective, I detect no evidence of hostility between the two species. They seem to know that the resources are plentiful enough to be shared. Even a sea lion seeking to steal a fish from the beak of a dolphin doesn't provoke a reaction. I watch until this interspecies feast ends, the last fish scales sink below the surface, and the dolphins and sea lions move off in separate directions.

Over the next months, I'm excited to observe these interspecies encounters with regularity. Where I had previously noted the presence of pinnipeds in our cetacean observations, I now gained new perspective born of following sea lions. Repeatedly, I see groups of sea lions seeking dolphins, which makes me wonder about the nuances of what gathers these different species of marine mammals together.

To shed light on this question, I decide to videotape all these aggregations, which I hope will enable me to analyze the behavioral sequences and gain insight into the details of this intermingling. I think one viable hypothesis is that sea lions may actually take advantage of the superior ability of dolphins to echolocate food. I decide to make it another priority of my research, especially considering I can find no long-term studies of sea lions and dolphins in company.

At night, if I am not too tired, I review and catalog the videotapes of aggregations we filmed that day. The process involves viewing the videos, often in slow motion to better understand the behavioral interactions, and entering these observations in a video log, which is tied to the time code of the tape. I don't often get seasick on the research boat. Over my years at sea, I've somehow learned to control it. But sitting in front of a video monitor, watching and rewatching these jerky images for weeks, could make anybody sick.

After slogging through stacks of sea-sickening videotapes and doing the data analyses on these interspecies associations, I'm finally able to prove my hypothesis. Sea lions do, in fact, seek out dolphins. They spend a significant amount of their time following cetacean schools, often spyhopping or leaping to keep the dolphins in view. Cleverly, sea lions invest time pursuing dolphins, which increases their chances of locating prey in the open ocean, where resources are patchy.

Logging more days at sea and studying associations between sea lions and dolphins, it's dawning on me that I won't really understand cetaceans in the wild without looking at what's around them, and identifying the linkages that exist between different species and their collective habitat. Everything is connected in the interweaving web of marine life: fish, seabirds, phyto- and zooplankton, kelp. Everything plays a part; everything is necessary. Seabirds are often present during our sightings, so it seems a natural step that they too should be a part of my data collection.

To help accomplish this, I've enlisted the help of ornithologist extraordinaire, Jon Feenstra. Jon comes from the Appalachian ridges of northwestern New Jersey, where he grew up following birds with a pair of binoculars semipermanently slung around his neck. He's in Los Angeles after finishing his PhD in chemical physics at Caltech. But it isn't physics that piques Jon's attention; it's birds: any species, in any form, anywhere around the world. These days, birdwatching is not only his passion but has taken over his professional life as well. His knowledge of birds is amazing; it seems Jon knows them all, by appearance, by the way they fly, by coloring or seasonal presence.

I've recently switched software for the dolphin project, and I'm now using the program Logger. The program allows me to automatically store data such as our GPS position, along with other oceanographic information from our instruments (water temperature, depth, wind speed and direction, etc.), all at predetermined intervals. My old forms are also integrated into a single data set, all georeferenced by time and position. The program is capable of storing different types of files that include digital images and video, as well as sound files we record through our hydrophones. This software, together with new geographic-information-system technology, allows me to re-create a sort of virtual representation of what was taking place above and below the water for any given sighting, affording a more comprehensive analysis than the methods I was using previously.

Because of the excellent results in shifting from clipboard to computer, I decide any seabird study should be digitally linked to the marine mammal data set. To accomplish this, I settle on a PDA running software that can be referenced to Logger through position and time. With Jon's help, I create a checklist comprising the most prevalent species of seabirds in our area. The data collection requires recording the number and types of birds found within three hundred meters and 180 degrees of a fixed position on the boat, as it proceeds along a predetermined transect. Jon continuously monitors this area during the entire survey—both in the absence and presence of marine mammals. As I did with sea lions, I begin looking at the relationship between seabirds and marine mammals, and the work takes on an almost three-dimensional aspect.

In an attempt to stay up-to-date with technological advancements, I buy a brand-new digital camera, and the need (and cost) of film is suddenly erased. Digital photography has revolutionized photo-ID, and I can now take pictures and preview them in the field, erasing what I don't like. Back in the lab, I'll download the good images to a laptop, sorted by date and sighting.

The downside to the LADP technological makeover is the cost. All this new gear is expensive, and the nonprofit cannot keep pace in raising funds. We've gotten some grants for our education and conservation programs, but they've been small and don't cover the research expenses. Our board members have not come through for us with the fund-raising efforts we'd hoped for, but we forge ahead anyway, trying to find new ideas to support our on-the-water work while raising public awareness . . .

We now have an active web presence and decide to try a ride-along program, where one or two guests from the general public come along on surveys to "spend a day in the company of dolphins," in exchange for a tax-deductible contribution to the nonprofit. The program is intended to be far more participatory than whale watching, in that guests can help with data collection under the guidance of an LADP researcher. This, we think, is a rare opportunity to learn about marine mammals and the problems that face them, and we hope that participants will come away with a connection to nature, perhaps providing some small basis for stewardship and respect for the ocean and its inhabitants.

Over the next months, the ride-along program brings a medley of personalities aboard. We are host to teachers wanting to learn more about cetaceans, Boy Scouts earning marine biology merit badges, octogenarian dolphin enthusiasts looking for video opportunities, wildlife photographers, energized birdwatchers, surfers, even a fashion designer. And most return to port with a different attitude than when they left. Perhaps it's proximity to nature, or maybe it's us rambling on for hours and hours about the environmental problems facing these animals, but it is unusual that our guests come home without some perceivable impact.

Something I've noticed time and time again is how nature changes people. It can peel off our outer layers, the insulation we all build up from our daily urban lives. It can lessen our need for all manner of things and help us see the beauty and magnificence of what is everywhere around us. I've seen it happen with my volunteers in the remote reserves of Yucatán, and I see it sometimes here, just a few miles seaward of the grand Los Angeles metropolis. I think most of us have some internal need to reconnect with nature, but we are so busy with everyday life that it slips by us, and we don't remember how much we need it.

I do what I do because I love it, but that love would never have developed had I not spent so much time in wild places with wild creatures. I continually ask myself how can I help others feel something of the love and respect for nature that drives me? How can I show them what I see, but through their own experience? The ride-along program, I think, might be a way to start breaking down the barriers between environmental awareness and personal engagement. It is a first small step toward changing things and protecting the critical habitats that we encroach upon daily.

12

Fingerprinted

"Oh my God! You study dolphins . . . How cool . . . !"

I can't begin to say how many times I've heard this. It seems that what I do is something many people dream about. The adventurous, romantic life of a marine biologist, out in the elements, investigating the lives of these magnificent creatures in the freedom of the vast ocean . . .

I am fortunate indeed, and I wouldn't exchange my life and career for anything. But people don't often realize what goes into the job. For every hour I log at sea, there are probably at least five to spend in the lab back on land. The work is as long and hard as it's rewarding, both on and off the water, but the many hours passed hunched over a desk as the clock ticks late into the night—analyzing, writing, correcting, rewriting—are where the less committed tend to weed themselves out of the vocation.

I look for research assistants and volunteers in a variety of places. Some come through our OCS website, other environmental nonprofit groups, or schools; others through friends and acquaintances. Only a small percentage of people, though, actually stay with the project through the six months of training it takes to become a competent LADP researcher.

As for me, I'm busy. Between meetings with my advisers, my responsibilities at OCS, the occasional seminar at UCLA, a two-day-a-week teaching job, and the research, I have little time for anything else. Sometimes it almost feels that my fieldwork gets in the way of these other obligations, and I have to remind myself that fieldwork is the reason I chose to do all of this in the first place.

When my schedule gets so hectic that I begin to experience a sort of detachment from the ocean wilderness, it is often the surveys that bring back my sanity. Sitting on the boat as it glides toward peaceful pelagic waters, I can relax and clear my head of statistical analysis, insecurities, funding uncertainties, and academic tribulations. This is what Richard Louv probably means by "the calming effect of nature" in his book *The Last Child in the Woods*. My "woods" are the ocean.

For one chapter of my PhD dissertation, I need to sort over 22,000 cetacean photographs along with the massive amount of ecological data that now clutters the floor of our apartment. The images need to be selected, labeled, and placed in chronological catalogs for identification of individual dolphins and whales. Matching the dorsal fins of bottlenose dolphins will be the biggest part of this undertaking, but I know it will provide me with valuable information on group structure, movement patterns, site fidelity, even population size.

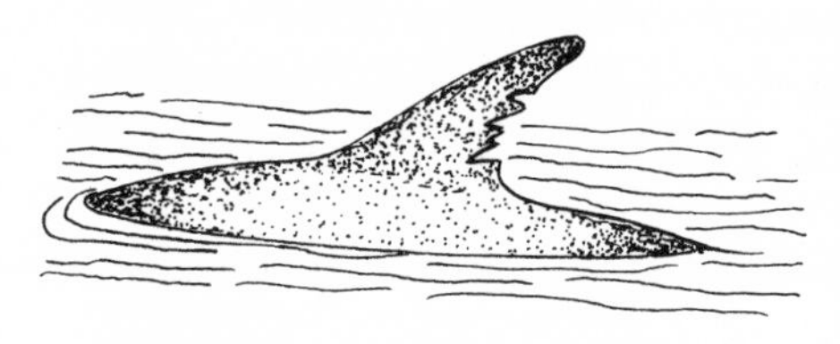

Fin matching is the core process of identifying dolphin individuals. This is the same system I taught my volunteers back in Greece, and it's a painfully slow and tedious process. A dolphin researcher friend of mine, however, points me toward a group of scientists at Texas A&M University who have just developed new software called Finscan, a computer-assisted identification system that automatically compares fin images. Finscan is still in the developmental stage, but it seems that it will trim weeks, if not months, off the previously used handmade fin-tracing process. The next four months are passed sorting and scanning over six years' worth of bottlenose fin photos . . .

I analyze almost 12,000 images of bottlenose dolphins. After a good amount of work scanning, matching, checking, and rechecking, I arrive at a total of 290 coastal dolphins, individually photo-identified by the long-term natural marks on their dorsal fins. These are my "metropolitan dolphins."

It isn't surprising to discover that coastal bottlenose is the most common species observed here. Nor am I surprised by the results that prove these coastal animals are not full-time residents of the bay, but remain in these waters only seasonally to forage and feed, sometimes only for a few weeks, before moving on, presumably north and south along the coastline of California and Mexico.

But one result of my research is unexpected. It shows these metropolitan dolphins are not so "metropolitan" after all. According to other scientists in California, coastal bottlenose dolphins remain less than one kilometer from shore, without venturing out into deeper waters.

My study illustrates that not only do coastal individuals travel near shore, but they also move along the escarpment and the submarine canyons farther off-shore. Every so often, they leave the narrow "coastal corridor" defined by other researchers. This is an interesting new finding, the significance of which I do not yet completely understand.

All fingerprinted dolphins in the bay are given a name. At the beginning, the names are based on some physical trait or behavior observed in the wild, like Half Moon, Milkspot, Bump, Big Notch, or Jump. As the analysis progresses, I run short of ideas, so I switch to more accessible names, like those of my researchers, their nicknames, the names and nicknames of their significant others, friends, and pets; even Burbank has several dolphins named after him. When these run out, I turn to Italian words my team can't even pronounce, like Scorpio, Gnomo, Linguetta, Striscia, Gobbo, or Mozza (short for mozzarella).

Melmom, Ascia, Cocco, Mogly, Duomom, Shelbemom, Ninamom, and Eos are among the most observed individuals in the metropolitan waters of Los Angeles. I saw them so frequently that I'm now able to recognize them at sea from their familiar fins. One year, I had the honor of changing Nina's name to Ninamom when I recorded her nursing her new calf. I saw her again and again after that for over a year, always swimming next to her offspring.

At one point, I was interested in trying to use bioacoustics to identify dolphins. Dolphins may have what are called "signature whistles," which are sounds specific to individuals, and once these are matched to an animal, the presence of that animal can be recognized by sound only. I seek the help of my old friend Fabrizio in setting up a system.

"You'll need to get a basic library of sounds for your area," he explains on the phone. "I can send you one of my assistants for a couple of weeks . . . to help you get going."

We already have some hydrophones, but he gives me a list of additional equipment we need to do the work. We have to record the sounds of different species on a digital recorder, after which we'll process the sounds using software called Raven. This will give us a graphic picture of each cetacean sound, called a sonogram. Armed with a library of these sounds, we can then begin to isolate bottlenose dolphin whistles.

Three weeks later, Michela arrives from ICRAM, the Istituto Centrale per la Ricerca Scientifica e Tecnologica Applicata al Mare, Rome, where Fabrizio heads the Bioacoustic Department. Michela works with Charlie making technical calibrations and tests, and after a few days of preparations, we head out to try the new system.

Our first bioacoustic survey doesn't go well. We follow our regular route, just off the beach, but all we hear are some unintelligible squeaks against the roar of the waves breaking on the shore.

"You need to go farther offshore," says Michela.

We try that too, but the sound of our own engines mixes with the dolphin sounds, and we are unable to get a clean sample. Charlie and Michela come up with the idea of mounting an electric motor on our boat for use when we are recording. This works fine during recording, but it doesn't push the boat fast enough to keep up with the dolphins.

On our first successful attempt, we are ten miles offshore, recording and photo-identifying a group of bottlenose. Every time the dolphins move away from us, Shana has to restart the outboard engines and turn off the electric motor, while we move close enough to continue recording. Charlie is on the roof of the boat, directing the recording. All in all, it's a mess.

Then something odd happens. All of a sudden, the dolphins disappear. A moment ago, I had them in my viewfinder, and we could clearly hear them "talking" in the headphones, but now there is no sound at all—and no movement. Charlie is holding on for dear life as the boat rolls back and forth with the swell. We are all wondering what happened as a slender, dark profile moves along our port side, just beneath the surface.

"Shark!" Charlie yells from the roof, and begins to hum the theme from *Jaws*.

The shadow passes under our boat again, and we can see it's almost as large as our boat. We are all leaning over the side to get a better look as the distinctive fin of a great white shark breaks the surface. Charlie tightens his grasp on the radar antenna. He stops humming.

For me, as a marine biologist, being up close and personal with such a rare and endangered animal is an amazing opportunity, and I start taking endless pictures of its silhouette. But as suddenly as it appeared, the shark vanishes into the blue.

The dolphins are long gone, and our bioacoustic experiment is over. After some further consultations with Fabrizio and a few more attempts to record dolphins, it becomes clear to me that I just don't have the time, skill, or resources to pursue bioacoustic identification at this juncture, and I go back to focus my attention solely on the photo-ID work.

Over a year passes, and I have spent most of my time in the lab putting the final touches on my PhD dissertation. It's finally time for my graduation from UCLA, and my parents fly out from Italy to attend. They are excited for me; at that time in Italy, doctoral degrees were fairly uncommon, and this is a big deal. Charlie plans a huge party at our apartment, spending three days cooking lasagna like a madman for the fiesta.

At my graduation, my adviser and mentor Dr. Hamner sits on stage, representing the biology department. After the ceremony, he comes outside to congratulate me and meet my parents.

"Oh . . . by the way, Maddalena," he says, smiling as he's leaving, "as I have told you many times, stop calling me Dr. Hamner . . . We're colleagues, call me Bill . . . If you don't, I won't talk to you anymore."

It was true that Bill had asked me often to be less formal and "European" with him during my stay at UCLA. I was the only PhD student he had who insisted on calling him Dr. Hamner. I suppose it was the social rigidity of European academia, so ingrained in my upbringing, that precluded any easy transition to the informality of American universities.

I am glad to be out of school again. It's an odd sensation, having spent many years inside a classroom. My phone rings.

"Maddalena, it's Bill. I know you've had some experience in curriculum development, and I'd like to offer you a postdoc fellowship to do some for me."

I tell him that my plans were to work full-time for OCS, expanding my research, but Bill assures me I can easily manage both. And there is a stipend . . .

"Easy for you to say," I think to myself, but I graciously accept his offer. Bill always had my best interests in mind, and this was no exception. And it wasn't really like going back to school . . .

My marine mammal study continues to evolve, as I work full-time for OCS.

I'm logging more time exploring the outer waters, sometimes venturing out a hundred miles or so. I'm following cetaceans for longer periods of time, often until darkness forces the end of the survey.

As Charlie becomes more involved in development of the research, his bias for going as far offshore as he can for as long as the rest of us can stand it begins to influence the LADP methodology. His interest in exploring pelagic waters, coupled with my newfound curiosity in the private lives of the still-unstudied offshore bottlenose, leads us to the idea of conducting multiday surveys, attempting to remain with a school of dolphins as long as possible. I have become increasingly intrigued by my finding that coastal bottlenose sometimes venture offshore, and I am almost convinced that the distinction between coastal and offshore animals might be less pronounced than other researchers had thought.

I formally increase my study area to include the entire Southern California Bight. Where I once brought only a single Pelican case on a survey, I now need a dolly to carry the stacks of cases full of my equipment. There is a camera case, housing two cameras with multiple lenses and accessories; a video case with a professional digital camera, a small HD camera, and sound equipment; another case that holds an underwater video camera, housing, cables, and monitor; and a case for hydrophones and a digital audiotape recorder. I have digital rangefinders, binoculars, even a headphone intercom so I can speak directly to the person taking data and shoot photo-ID at the same time. Crew size also grew, to where I rarely leave the dock with fewer than six people aboard.

At times, I think about how much my work has changed since my early days with lizards. In a way, I miss the simplicity of my old notepad. All this accumulated equipment and technology adds new layers and possibilities to the research, and it helps provide a more comprehensive view of what I see in the field. But I have taken great care to make sure that technology does not take away from what is important to observe at sea. At the end of the day, I want my work to essentially remain as it was in the beginning: based solely on an open-minded observation of animal behavior.

Unfortunately, the cost of gasoline has increased greatly as well, and after a few good years of use, I realize our Pro-Line, with its two large outboard engines, isn't well suited for my fieldwork anymore. To replace it, we purchase *Gone with the Wind*, a used fifty-foot Santa Cruz racing sailboat with lots of space to work, crew quarters for eight, a full galley, and a small, economical diesel engine, in addition to its ample inventory of sails. Charlie invests time making all the instruments and computers aboard talk to each other, so Logger can acquire a broad variety of oceanographic and position data.

Gone with the Wind is the ideal vessel for offshore surveys. It motors comfortably at nine knots, it's quiet around dolphins, it offers a great working platform, and, weather permitting, we can use the sails for propulsion at zero cost. And sailing is nothing short of awesome on this boat. She is fast and stable and can easily reach powerboat speeds under sail alone. The wheel is almost as tall as I am, and when driving, I feel the full power of the boat, the wind, and the waves. I now understand why my husband loves racing so much. The feeling of speed, driven only by the wind, is exhilarating.

But working on a thoroughbred sailboat like this one was different from working on a small powerboat. My crew needed to learn seamanship and some essential sailing skills along with their research duties . . .

"The rabbit comes out of the hole, runs around the tree, and goes back down the hole," Charlie recites, slowly using a piece of line to demonstrate how to tie a simple bowline. But once he passes the line to my assistants, the rabbit never seems to get where it has to go. It appears good researchers don't necessarily translate into good sailors . . . Eventually, everyone gets it, and we head out, with the wind behind us.

One year of working on *Gone with the Wind* provides me with enough data to compare the photo-ID and behavior of offshore bottlenose dolphins to that of their coastal counterparts. While the behavior for the two populations seems to be different, with offshore dolphins socializing more than coastal dolphins, the photo-ID analysis confirms my belief that the spatial separation between coastal and offshore animals is not so pronounced. The only way to know what is really taking place is to collect more information, then compare my data with the work of other scientists along the California and Mexican coast. And I need money to do so.

Through OCS, Charlie and I have been busy trying to find funding for the research. Our work finally pays off when we receive a call from the director of the Santa Monica Bay Restoration Commission (SMBRC), who is interested in funding a long-term monitoring project on marine mammals in the bay. After much deliberation, our funding is approved, and the first order of business is to broaden our baseline knowledge of coastal animals, collaborating with other scientists along the Mexican and California coast to create the first comprehensive "fingerprinting" of California coastal bottlenose dolphins.

Things are going well. I am now Dr. Maddalena with a postdoc, not that it makes much of a difference. But I'm not a student anymore, and that is a good thing! Charlie and I are working together full-time at OCS, as we had hoped to.

We are doing at least two surveys a week and are branching out toward more of a conservation orientation for OCS. We even start an adopt-a-dolphin program through our website, where anyone can help support our on-the-water research by "adopting" one of the individuals we have identified, with a donation. We prepare an adoption kit that contains photos of the specific dolphin, a sighting record of when the dolphin has been seen, a comprehensive cetacean booklet, and a personalized certificate.

"Hello, is this the Ocean *Conversation* Society?" asks the voice on the phone.

"*Conservation* . . . yes, this is the Ocean Conservation Society," I reply.

"I saw your website and I want to adopt Mozza," says the woman. "I live in Dallas and I really want to help out."

"That's great, we need all the help we can get."

"But I have a question," she goes on. "I have a swimming pool in the backyard, but is there anything I need to do for the adoption? I mean . . . do I have to put salt in the water or something? And how does the dolphin get here?"

When I gently explain that the adoptions are more of a figurative rather than physical thing, the phone goes dead. Probably for the best.

PART 3

Hard Times

13

Invading Their World

The rotting carcass of a twenty-foot gray whale floats two miles off Manhattan Beach pier. The flukes are lassoed with a heavy rope tied to several buoys, suspending the body at the surface. A dozen western and California gulls tear at the dead cetacean, flying off with chunks of meat in their bills.

"What are we going to do?" Karyn asks me, as we approach the whale.

"I'm calling John at the museum," I say, keying John Heyning's number on my cellular phone. I'm sure he already knows about this, but it won't hurt to double-check. As we pull alongside the whale, we can see its skin has turned a yellowish-brown, and there are long, tangled strands of derelict fishing line wrapped all around its head and flippers. It's a heartbreaking sight. This animal probably suffocated when the fishing line made it impossible for it to swim to the surface to breathe, or it died of exhaustion.

Another gray whale was recently found dead on the beach of Elliott Bay, in Puget Sound. In the necropsy performed on this thirty-seven-foot male, scientists discovered an astonish-

ing amount of garbage in his stomach. They found more than twenty supermarket plastic bags, a pair of surgical gloves, duct tape, a pair of sweat pants, a few towels, a golf ball, and a variety of other plastic items. The poor whale had half the contents of a convenience store in his belly.

My home port of Marina del Rey has a detached breakwater that protects the harbor from the ocean swells. The breakwater is a gathering place for seabirds and pinnipeds alike, which sometimes number in the hundreds. On many occasions, I have witnessed the direct effects of derelict fishing gear on these animals. Sometimes it's just small strands of line that trail from the feet or feathers of gulls; other times it's a large hook protruding from the bill of a pelican or the ugly open gash in the neck of a sea lion where entangled fishing line cuts continually deeper as the animal grows in size, often ending in death. The sad part of these encounters is that they are not isolated incidents.

A week after seeing the gray whale carcass, I meet John at the marine mammal warehouse. We plan to talk about new developments in my research, but the conversation quickly slips into an account of the sighting of the decomposing whale and the effects of marine debris on marine mammals. In John's opinion, debris is frequently responsible for the death of these animals, even here, in our ocean "backyard."

Leaving the warehouse, I think about what I've seen over the past several years in the bay: all the Styrofoam plates and cups and the endless rivers of plastic bags and "Happy Birthday!" balloons floating on the surface. With balloons, it was almost a joke at the outset of my fieldwork. We spotted and collected balloon bouquets at sea almost every time we went out on a survey. Burbank was our official deflation engineer and the only one aboard who looked forward to the experience, with tail wagging as he attacked the balloons, ripping them to pieces.

But the joke became more of a depressing routine. We couldn't travel more than a few miles without stopping and picking up trash from the ocean, and in the course of a dolphin survey, we could easily fill two trash cans with the wayward plastic. As much as we were marine biologists, we were ocean janitors.

I'm sad knowing that today dolphins and whales share the ocean with staggering amounts of marine debris. Trash, mainly plastics, is responsible for suffocating and drowning over 100,000 marine mammals and millions of seabirds in the North Pacific every year. There is no place left in the ocean where litter cannot be found, mostly because it can last for centuries in the form of minute, raw plastic pellets. There is no safe place for cetaceans, or any other marine creature for that matter. And worst of all, the average density and composition of debris have continued to grow at a frightening pace over the past decades.

Plastic doesn't stay in one place at sea, as one might think. It may travel for thousands of miles, from the waters of the Atlantic Ocean to cold Antarctic circumpolar currents. The Great Pacific Garbage Patch is no joke. It is a scary reality of modern times: a colossal gyre of marine debris in the North Pacific Ocean, estimated by some to be twice the size of Texas. Wind-driven and captured by currents, a high concentration of suspended synthetic and semi-synthetic industrial products make up this massive field of pellets, which are gradually drawn toward the center of the gyre. It's difficult to see on the surface, but just a few inches beneath lurks the ghost of several million tons of plastic, like something out of a science-fiction movie. Humanity uses around 300 million tons of plastics each year. Do any of us ever wonder where all this plastic might end up?

I was kayaking with Charlie on a blustery afternoon in Marina del Rey. We paddled out to the mouth of the harbor in a strong onshore twenty-five-knot breeze to play in the building waves preceding the first major storm of the season. We reached the harbor entrance just as the storm came ashore, letting loose with a torrential downpour. It was truly spectacular.

Then something unexpected happened. Ballona Creek—immediately adjacent to the marina entrance, which serves as a major storm-drain outflow for the city of Los Angeles—let loose with a vengeance. All of the trash, debris, oil, and gasoline that had accumulated on the streets of the city over the rainless summer now washed out to sea in an amazing torrent of coffee-colored, reeking water. It was as though we had been hit with a tsunami of disgusting trashy water, and it was all we could do to keep from being swept out to sea in its fury.

With Charlie's help, I negotiated the six-foot standing waves of plastic bags and garbage, dodging king-sized mattresses, car seats, whole sheets of plywood, and all manner of urban junk. Somewhat miraculously, we made it back into the harbor entrance, completely covered in debris and stinking of who-knows-what. Remarkably, neither of us became ill from the experience. The following day, I checked satellite images of the area and found that the debris and pollution plume from Ballona Creek extended for miles out to sea!

A few years ago, I took a trip down the Pacific coast of Baja in search of coastal bottlenose dolphins. After driving for an hour on a dirt road to an isolated beach, I came upon what would have been a magnificent stretch of shoreline, were it not for the trash dump that extended as far as I could see in either direction. Instead of driftwood, there were bottles and fishing floats washed ashore, and the beach was littered with dumped beer cans, condoms, refrigerators, and all forms of unwanted stuff. A northwesterly wind was blowing wildly, lifting objects from the ground and scattering them across the adjacent dunes in random order. As I retreated toward my car to avoid the rain of garbage, I could see gulls passing by and picking up trash, then flying out toward the open water. It was a sad sight.

We still believe the ocean is an infinite dump that will somehow absorb whatever we throw in it. Even in my own neighborhood, where environmental awareness is prevalent, our beach sand is intermingled with bits of Styrofoam, glass, and Mylar food wrappers, so much so that we hardly notice it. We are so accustomed to living with our waste, it's becoming an integral part of our natural scenery. We have traded plastic and cigarette butts for the shells and starfish once brought in by the sea.

We are good at pointing fingers at each other or at big corporations, without accepting any personal responsibility. But strange as it may seem, pollution from marine debris is largely caused by the unintentional acts of large numbers of people who, individually, toss trash out into the world. Throwing a cigarette butt or a six-pack ring out the car window might seem an insignificant and isolated action, but it's the cumulative outcome of all those actions that is suffocating the oceans. This, coupled with the incessant and growing production of synthetic materials that do not easily break down in the environment, contributes to the devastating biological impacts of marine debris.

We are floating at the edge of a dense kelp bed near the Palos Verdes peninsula, waiting for the coastal bottlenose dolphins we've been following to move a little farther offshore so we can finish our photo-ID work. We identified two of them, Cocco and Ascia, just before the school dove deep under the canopy of the sea forest, only to resurface on the inshore side of the kelp. In no hurry, we munch pretzels and talk about life while Charlie tends the wheel, pacing the slow movements of the dolphins.

"Pretty amazing sightings this morning," says Charlie.

Earlier that day we had seen four thresher sharks in succession, suddenly leaping clear of the water directly in front of our boat. They were young, around three or four meters long, and none of us had ever seen them breach like that. We watched as they slapped the water with their tails, probably to stun prey. It was fascinating!

Ascia, Cocco, and company swim out of the kelp bed, reappearing near some lobster traps several hundred meters from us. They turn west to follow a front, as if they were taking a well-marked freeway entrance. A front is the interface between two water masses of different temperature, and where fronts exist, I often find aggregations of a diverse array of marine creatures, from tiny zooplankton to marine mammals to seabirds. Many species use these fronts to forage and feed. And it's here, in these convergence zones, that we also come across the highest concentrations of marine debris.

As we enter the frontal zone, two young bottlenose approach the boat and begin bow riding. Aided by our bow wave, the pair dances and plays with each other near the surface, seemingly unaffected by the debris that surrounds them. Abruptly, one dolphin snatches a plastic straw, keeping it straight in its beak like a cigarette. The other casts a glance at its "smoking" companion, then bites down on a Styrofoam cup, breaking it in two pieces.

Astonished, I follow the two juvenile dolphins that continue to cavort with each other, playing with an assortment of debris, as if they were toys. I see the "smoking" dolphin drop its straw and grab a large piece of a plastic bag in its mouth. The bag disappears, but I never see the animal release it.

"Are these guys eating plastic?" I yell to Charlie, who is videotaping the whole thing.

"I can't tell," he yells back. "Looks like it . . ."

The dolphin snatches another plastic object, and neither of us see it let go of that either. What initially looked like dolphin play is now turning into a more treacherous game. It appears that at least one dolphin has swallowed its plastic toy.

Aside from the danger of suffocation, several species of marine mammals can die of starvation when eating plastic debris gives them the false feeling of filling their stomachs. Marine litter can also become highly toxic as it "absorbs" pollutants such as DDT, PCBs, and chemical compounds from the seawater, causing ill effects that may be difficult to isolate.

We pass into a regatta of thousands of by-the-wind sailors that blanket the entire width of the front. These are free-floating jellyfish of the genus *Velella*, painted in stunning blue shades, the largest of which is barely the size of my little finger. They have tiny sails to catch the wind, which propels them through the water like a sailboat. They are probably brought here by currents along with plankton, their preferred food, which they capture using their nematocysts, the minute toxin-filled harpoons that they launch from the short tentacles dangling under their "skirt."

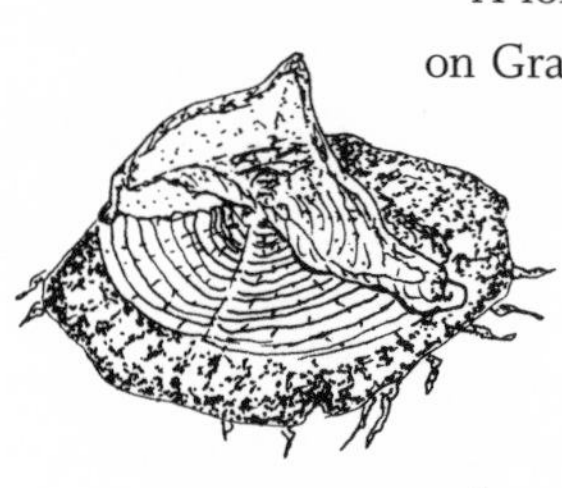

A long time ago, I saw these jellyfish on an isolated beach on Grand Canary island. This is the first time I have ever recorded them here. I dip my hand in the water, and a *Velella* lands on my palm. It doesn't sting and looks less elegant out of its element. I show the *Velella* to my crew and immediately drop it back into the water to continue its journey with the others at the mercy of the wind.

Between dolphins, acrobatic thresher sharks, and *Velella*, it's been a day filled with unusual sightings, but on the way home, the image of the two dolphins playing with trash sticks in my mind. I increasingly see the effects of irresponsible human consumption, and I am growing angry over our collective lack of stewardship and respect for the oceans.

In the first couple of years at OCS, we made presentations to schools, worked with other nonprofits, and developed educational programs designed to raise public awareness of the environmental problems facing marine life. We promoted recycling programs, kayak-based cleanups, and anti-littering campaigns. But we began to notice that, even though public awareness in our area was on the rise, there were few firsthand, action-oriented opportunities for people wanting to make a difference. In our presentations, we were having an increasingly difficult time explaining how recycling would "change the world," especially given that the problems of ocean debris were worsening on a global scale. We needed a hands-on approach that would provide not only an educational experience but also a feeling of accomplishment for the participants and a means to make a measurable impact, at least at the local level. But what?

Several of us at OCS were doing a presentation on marine conservation at a Los Angeles high school for a group of science honor students from the inner city. Driving there, I was struck by how different their childhood surroundings are from those where I was raised. These kids came from a community torn by gang-related violence and drug abuse, a community mostly ignored when it comes to the distribution of educational resources. Many houses had security bars on the windows and doors, and the main thoroughfares were covered by gang graffiti. The school had put up tarps on the perimeter fences to guard against drive-by shootings, and the students (and I) had to pass through metal detectors at the front door.

Our presentation focused on the adverse effects of urban runoff on the marine environment, but as I gave my speech, I found myself wondering why anyone who had to deal with this reality on a day-to-day basis would care about the health of the ocean. Many of these kids had never even seen the ocean.

On the way home, Charlie and I talked about this lack of relevance and how we might find a way around it. The reason we care as much as we do about the health of the oceans is because we have both cultivated a relationship with the sea, albeit from different paths. We care to protect the ocean because we love it for its intrinsic vastness and beauty, for its biodiversity, and for the experiences it provides us. The challenge, we decided, was to find a way to impart this connection to other people so that they, too, would care enough to act.

From this discussion, the Conservation Mentorship Program was born. Students with a strong interest in marine biology, oceanography, or conservation were selected from different schools based on their merits and personal interests to undergo a comprehensive training program specifically designed by us. The program provided not only the basis of marine mammal field research, but also a foundation in environmental conservation. We did this by providing semester-long participatory research opportunities, coupled with a community outreach component and a final project that would have a measurable impact on a local environmental problem. This was different from writing an essay or taking a multiple-choice exam; it required conceptualization and action to implement. Our hope was that the program would teach students how to bring an idea to reality, something their regular classes did not usually accomplish.

Over the years, OCS sponsored and ran numerous mentorships, but the program eventually failed. Not for lack of funding or because it didn't work: in fact, everyone loved it and called it a tremendous success. It failed because of the resistance we encountered from school administrators. It turned out that something extracurricular, as this program was, made more logistical and coordination work for teachers and administrators, and they simply did not have the time or inclination to deal with it.

I sincerely hope that some of the many young students we worked with in the mentorship program will be environmentally active in their lives. Perhaps some will even join the next generation of environmental activists and policy makers. I loved the program because, for once, it allowed me to see something concrete being done. It provided more than just words and concepts.

Sometimes, when we go out on a dolphin survey, we can spend all day cruising among an endless amount of plastic bags and other urban debris. On those days, I feel it is hard to make a difference, even a tiny one, and I feel powerless in the fight to maintain a livable environment for the animals I study. We have invaded their world; in every corner of the earth, in the deepest ocean abyss, there is the obvious footprint of humanity. On those days, I need to pull myself together. I try to remember the words of the Dalai Lama, "Real change must start with individuals," and begin thinking about new approaches to the problems that, in some small way, may help return the ocean to its rightful inhabitants.

14

Dolphins Under the Weather

I am sitting on the dock in front of my apartment, watching a handful of sea lions resting under a mildly hot sun that just managed to burn its way through the thickness of the morning fog. For a couple of hours last night, the sea lions kept me awake with their loud barking, but now they are quietly relaxing, piled on top of each other like a confused mound of brown fur coats.

This morning, I need to get out of my office. My head is so clogged with the distressing events of the last weeks I can't work on my writing. As I watch the smallest sea lion raise its head for what seems a long and liberating yawn, images scroll through my brain . . .

The dawn is muggy and windless, as a bottlenose cuts through the calm coastal waters of the Gulf of Mexico. The dolphin surfaces to exhale, but what comes out its blowhole is not the normal respiratory waste one usually sees: it is a mixture of air, mucus, and oil! Oil . . . ?

During my writing of this book, nearly 180 million gallons of oil spewed into the Gulf of Mexico due to the explosion of a BP oil-drilling rig that killed eleven workers. The daily spill rate of this petroleum hydrocarbon was about 60,000 barrels—or 2.5 million gallons. This means that somewhere around 25,000–35,000 gallons of crude oil polluted the Gulf in the time it takes a reader to finish this chapter.

Something like 4,000 seabirds, 500 endangered sea turtles, and more than 70 marine mammals have died since the disaster first struck. And while some of the brightest minds worked toward the most advanced technological solutions in an effort to put the brakes on this environmental nightmare, things just kept looking dimmer. It is the worst environmental disaster in US history.

What really scares me is not so much the images of pitiful, oil-soaked pelicans or the economic impact on the shrimp crop; the truly frightening part is what will happen now that all evidence of the chocolate mousse–looking slicks has disappeared, and people have gone back to their everyday routines. We and the entirety of marine life in the Gulf will be feeling the consequences of our reckless oil addiction for years to come, and we have little understanding of how these consequences will manifest themselves over time. But what dominated world news for months is now gone from our televisions and our memories, and I wonder, *Just how blind are we?* Quite blind enough, I think, to repeat the disaster in the name of economic growth.

I reflect on the *Exxon Valdez* catastrophe, which today appears like a tiny accident compared to the magnitude of the BP spill. Twenty years after that ecological disaster, 21,000 gallons of oil still lurk beneath the surface of Alaska's Prince William Sound, exceeding all predictions, and continuing to affect sea life even today. But nobody talks about this anymore. It's a problem of the past. We learn little from our experiences, and I wonder how bad things need to get before we open our eyes, remember the past, and do something for the future of the environment we all live in.

The little sea lion is now awake and wriggles out of the bulky pile of sea lion bodies to take a swim. A couple of its companions voice their complaints by emitting several uncouth yelps, but they soon resume their positions in the cozy mass. Another youngster adjusts its head on the belly of a large male, as though it was a comfortable pillow. Now they are all back asleep.

As I watch the sea lions, my thoughts now drift back several years to one conversation I had with John Heyning at the Natural History Museum. We talked about oil dependency and the resulting risks to marine life. As my studies progressed, the topics of our meetings slipped more and more away from pure research toward discussions of the environmental challenges facing cetaceans. John had recently been promoted to deputy director at the museum, and things for both of us were going well. That day I could not have known that it would be the last conversation that John and I would have. Just after that encounter, John's dynamic life began to slip away from him, and after a painful and devastating battle with Lou Gehrig's disease, he died at age fifty.

At John's funeral service, his family asked that we all dress in Hawaiian attire; they passed out flower leis at the door to the sound of a Jimmy Buffett CD. The North American Mammals Hall at the museum virtually overflowed with family, friends, and colleagues who loved and respected John.

I would miss his colorful obsession with Hawaiian shirts, his profound devotion to the oceans and its inhabitants, his moral and scientific support, and even his jokes (though I didn't always understand all of them). John once gave me a retouched image of a cow and a dolphin leaping out of the water next to each other, to help illustrate to my students at UCLA that dolphins come from ungulate ancestry . . . John was a character, an inspiration, a great scientist, and a friend. I miss him to this day.

My sad reverie is interrupted as a small powerboat with eight people crowded aboard moves close to the dock where the sea lions are at rest. The passengers are screaming, trying to get a reaction from the animals, but there is no response. The driver makes a quick turn with the boat, throwing a violent splash that douses the pile of sleeping sea lions. The sea lions are awake now, swinging their heads back and forth in a clear sign of disapproval and barking at the intruders. The people on the boat are laughing vociferously, and one of the kids aboard makes a ball of some trash and throws it at one of the pinnipeds. I can't watch this scene any longer and get up from where I am seated.

"Hey," I say, upset, "you can't do that!"

They see me, and one kid mocks my accent: "You can do dat!" he screams back.

As the driver speeds up to drive away, the kid gives me the finger.

The sea lions are gone, diving into the water that, today, has a particularly ugly look, probably from another gasoline or diesel spill. These small spills happen so frequently in the marina that the Harbor Department tends to overlook all but the worst of them. I watch as the streaks of dispersing fuel draw intricate swirls on the water surface. Not only do sea lions have to deal with human harassment, but they live out their lives in these polluted urban waters as well. California sea lions have one of the highest concentrations of DDT and PCBs found in marine mammals.

Production of the dangerous DDT was banned in the United States in 1972, but not before hundreds of tons of the pesticide were flushed into Santa Monica

Bay by the Montrose Chemical plant just ten miles south of where I live. Montrose had been manufacturing DDT since the 1940s, and the resulting deposit of this pesticide is one of the largest in the world. It now rests on the seafloor in a toxic, gelatinous soup extending over seventeen square miles at the south end of the bay. These pollutants stay in the marine environment, contaminating every creature exposed to them, including dolphins and us. Forty years after Montrose, these waters still have high levels of pesticides.

Dolphins are top predators, meaning they feed at the top of the food chain. When chemicals settle into seafloor sediments, they are absorbed by a variety of small organisms. Some of these creatures end up in the stomachs of bottom feeders, which in turn accumulate higher concentrations of the same contaminants in their body tissues. Every time the contaminants move up the food chain into a new predator, the concentration intensifies in a process called biomagnification. By the time the contaminants reach the adult dolphin population, the concentrations are severe—so much so that stranded dead dolphins are currently handled and disposed of as hazardous waste. Pollutants also pass from one generation to the next. Through milk, dolphin mothers transfer sublethal doses of harmful chemicals to newborns during a lactation period that can last up to two years.

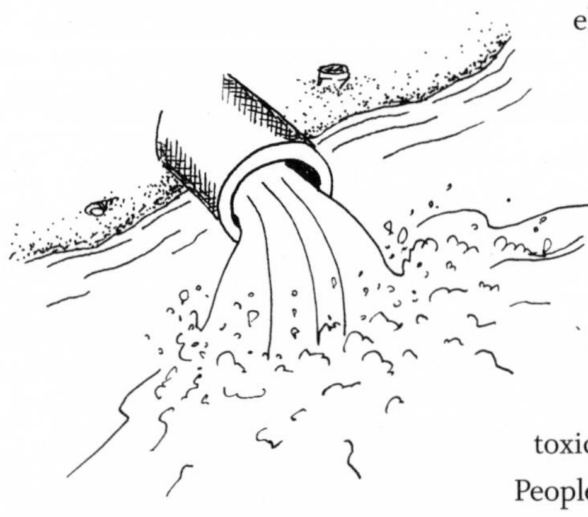

I often see dolphins and sea lions traveling in a liquid mixture of trash and the oily remnants of urban runoff. Even in the face of increased environmental regulation and improved source control systems, dolphins still swim in the hundreds of pollutants that wash down the city's storm drains, including all sorts of heavy metals and toxic compounds.

People tend to project human attributes on dolphins. We see them as carefree and joyful creatures that spend their lives cavorting in the waves, perhaps because of what we perceive to be a dolphin's ever-present smile. But I can't begin to count the times I've logged hours collecting data on dolphins feeding near a storm-drain outflow, in water the color of dirt. And while they superficially appeared "happy" and well to me, it didn't mean they were not suffering the effects of exposure to pollutants.

This was something I couldn't see at the inception of my research. Only time and baseline data would reveal the truth. The more time I spent following dolphins in the wild, then scrolling through thousands of images of dolphin dorsal fins and bodies, the more I began to recognize that the creatures I was studying weren't doing so well after all. I noticed signs of poor health in the form of lesions, spots, and open sores on their bodies. I saw physical deformities like hunched backs, grotesque protuberances, and spinal malformations. Nearly 80 percent of photo-identified individuals exhibited at least one type of dermal lesion; many had multiple types. As I looked deeper into the possible causes, I found that bacterial, viral, and/or fungal infections were at the top of the list, possibly correlated to bad water quality or contaminated prey animals.

Up to then, there hadn't been many studies on skin diseases and physical deformities in dolphins, and nothing at all had been done on the West Coast of the United States. The investigations conducted elsewhere suggested that these issues might be human-induced, likely in relation to poor water quality. Surfing at Venice Beach or Malibu for an hour after a recent rainfall could literally make me sick. And dolphins are apex predators, roving all day long in the very same waters. In addition, there are red tides, some of which are considered harmful algal blooms. These involve a species of phytoplankton that can be toxic to shellfish and other sea creatures and may cause severe respiratory problems in humans. We still don't know whether these harmful blooms are a natural occurrence, but there is increasing evidence, in some areas, of a link between their frequency and high nutrient loading caused by human activities.

One may care less about dolphins or want nothing to do with nature, like Woody Allen, who described his aversion to the outdoors by saying, "Nature and I are two." I understand that there are different points of view on this subject, and I'm aware that biophilia (the love of nature) and biophobia (the fear of nature) both exist. But it really doesn't matter what any of us think. Dolphins are barometers of how our oceans are doing; if dolphins are sick, it means the ecosystem is probably sick, and we may get sick too.

Not too long ago, I took an early morning walk with Burbank on the Venice Pier. In the dawn silence, a group of Latino fishermen were casting their lines into the water.

"*Otra corvineta blanca!*" one of them shouted, tossing a large brownish-white fish into a bucket.

I walked by and saw a dozen *corvinetas* still thrashing around in his pail. These are bottom-feeding fish known as white croaker, one of the many species of fish common to this area that dolphins often eat. This man would likely take the *corvinetas* home to feed his family. As I passed, I wondered if I should say something, if I should tell him this is one of the most contaminated fish species in the bay, with the highest concentration of DDT. As I was about to speak, the fisherman turned toward his friends, proudly showing off his catch. I decided not to say anything, but while I headed back toward the beach, I felt guilty for doing nothing.

A few days later, my skin-disease publication on dolphins is published online, and I wake up to find my e-mail in-box jammed with messages. They are from researchers from all over the world, asking me questions, sending pictures of dolphins in pitiful condition, far worse than what I've witnessed in my area. They too are seeing similar problems in their respective backyards.

A month passes after the day I spent sitting on the dock with the sea lions. Charlie and I have emptied the contents of our apartment, stacking our belongings in boxes in a storage locker. We are ready to leave. The car is packed with personal and technical gear, our kayaks are strapped on the roof rack, and Burbank is asleep on his blanket in the back seat of our 4Runner, curled comfortably around a brand-new stuffed bunny we purchased especially for the trip. Destination: Labrador! We are taking Burbank to visit his "homeland." Charlie promised to take him there when Burbank was just a puppy. Our dog is now eleven years old, and time is running out.

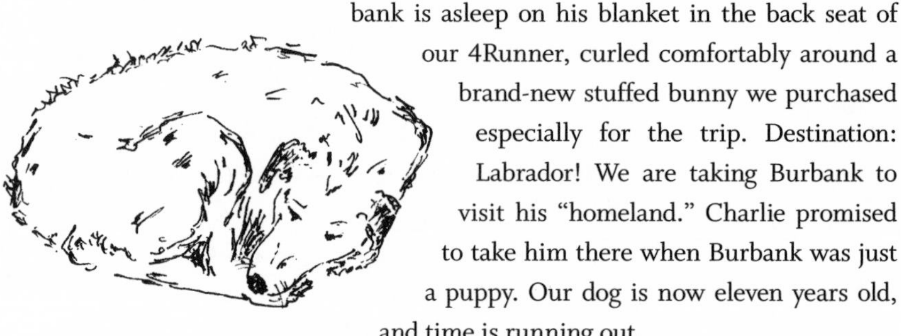

The dog's age aside, it's the right time to leave. Our research grant from SMBRC dried up after just two years of marine mammal monitoring. We completed a 200-page report along with several peer-reviewed scientific publications on our work, which was supposed to be ongoing. But suddenly the commission stopped supporting the monitoring of marine mammals, and we were left with no money to carry on the project.

These are hard times for everyone, not just dolphins. The country's economy is worsening, and public funding is getting scarcer by the day. After over a decade of dolphin research in California, we find ourselves in need of "downsizing" and looking elsewhere to find new and creative ways of supporting OCS and the research. Except for Charlie's Aunt Marcy and a few friends, we are not necessarily tied to California.

Sometimes my life feels like a roller coaster. Working at OCS, I am essentially a freelance marine biologist. I usually know what I'll be doing tomorrow, at times even for the next six months or year, but not often longer than that. The job I chose, that of working in a small nonprofit, is hardly stable. On occasion, I wonder about my future, what I will do when I grow old, where I will live, how I will pay for my existence. I miss my family and friends back in Italy, and I miss the lifestyle and culture of my country. I think about the beauty of the Italian countryside, of the small roads and towns and the flavors and smells of the Mediterranean. But, like all things, there is another side of the spectrum . . . I'm passionate about the animals I study, and I don't have to wake up every morning and go to an office or spend hours in traffic. I can be next to the person I love, working together on new and exciting ideas and projects for the nonprofit, and I have fun doing it. I get to travel to remote natural places where tourists never go. I suppose it all balances out in the end . . .

With OCS, we would like to continue the research and our educational outreach work, pushing the latter even further toward action-oriented environmental efforts, but we don't want to be pulled in directions we don't want go, just because there is funding available. We stopped writing curricula for that reason. There was plenty of money for curriculum development, probably enough to cover all OCS administrative expenses and then some. But we didn't think there was a real need for what we—and many other organizations for that matter—were producing. In fact, much of what we created never got through the bureaucratic process to the students for whom it was intended. Before we knew better, we actually wrote that in the final report to one of our grantors. We thought that public money was being wasted creating materials that most teachers had no time to read, let alone implement. We proposed instead that money be spent on finding a way to get already existing curricula to students, rather than on the never-ceasing creation of new materials. Clearly, they didn't like hearing this, and we never got another dime out of them.

When we founded OCS, we wanted the nonprofit to grow large enough to support itself, but remain small enough to avoid falling into the vortex of bureaucracy and compromise. We tried never to lose track of what our initial goals were and stay focused on things that made some measurable impact. Perhaps it was a naïve view, but it was, and still is, what we believe in. But without grant money, we found ourselves needing some time to refocus.

Charlie and I think traveling might give us a chance to clear our minds, as well as to explore other opportunities and potential collaborations for OCS. I have contacted several scientists on the east coast of the United States and Canada, and we plan to go there to meet and talk to them about future projects.

In Fort Pierce, Florida, I set up an appointment with Dr. Edith "Edie" Widder, deep-sea explorer and marine biologist. We corresponded several times about a potential partnership between our two nonprofits. The idea is to deploy a series of the football-sized water-quality sensors she invented, called Kilroys, to track contaminants back in my study area in California. This is something I'm particularly interested in, especially after the latest findings of my study on skin diseases.

We find a dog-friendly hotel in Fort Pierce the evening before our meeting with Edie. Charlie, Burbank, and I walk to the beach to take a dip into the lukewarm waters of the Atlantic Ocean to rid ourselves of the sticky Floridian humidity and escape the throngs of mosquitoes buzzing around us. The next morning we all head over to Edie's office.

"Welcome, Maddalena!" Edie says, shaking my already sweating hand. "Come upstairs so we can talk for a while with some air-conditioning on . . ."

I walk to the second floor of the Duerr Laboratory for Marine Conservation, winding my way between refrigerators, computers, stacks of papers, and Kilroy prototypes. We chat about our respective research, but the conversation quickly shifts to a discussion of persistent organic pollutants and endocrine disruptors, compounds that defy environmental degradation and tend to bioaccumulate in dolphins. The latter are agricultural and industrial compounds like PCBs that, once ingested by an animal, behave like hormones in their endocrine system, damaging regular physiological functions. Scientists have become increasingly concerned about the effects of these chemicals in recent years, particularly in dolphins, where these substances are known to disrupt fertility, sexual development, and behavior.

"Do you know about the dolphins in the Indian River Lagoon?" Edie asks.

"I've heard about them," I reply, "but I've never seen them."

I had read that a goodly percentage of bottlenose dolphins in the lagoon were affected by lobomycosis, a mycotic infection also found in humans, and by oral and genital tumors, of which I'd seen some grotesque images online. Further, many dolphins had pneumonia, hepatitis, meningitis, and disorders of the nervous system. But I had no clue how dire the problem had become in the last few years.

Saying good-bye to Edie, we promise to stay in touch to collaborate on writing a grant for Kilroys in California. While we are there, I decide to go and check the dolphins in the lagoon for myself.

The next morning, Charlie and I slide our kayaks into the murky waters of the Indian River Lagoon. Burbank positions himself comfortably on Charlie's boat, as he's done many times before. We start paddling, and there is no question from the smell and appearance that these are polluted waters. Freshwater discharges degrade water quality on an everyday basis, and the place is fertile ground for harmful algal blooms to flourish. As we paddle through the unmoving waters of this shallow estuarine ecosystem, I try unsuccessfully to imagine how this place might have been before industrialized agriculture came along and ruined it.

We meet a couple of bottlenose, after kayaking almost three hours under a blistering sun, and I can't believe what I see. The larger of the two dolphins surfaces next to me as the other moves away and disappears. It's dreadful. It is as though this animal emerged from a horror movie. Grayish-white abnormal growths cover almost half of its ulcerated body. Its skin has the consistency of cauliflower. The disfigured dolphin glances at me, pauses for a moment, then dives. And that is all I see of it.

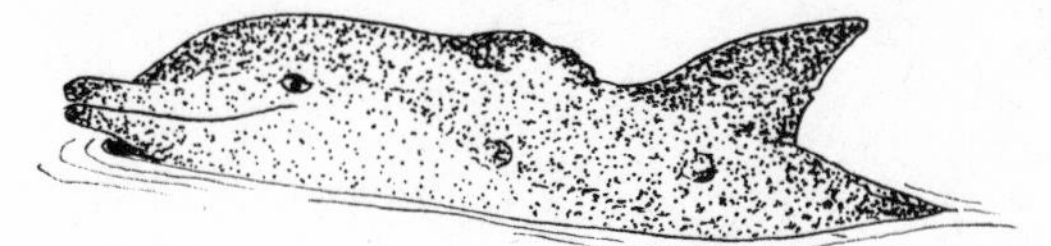

15

Vanishing Innocents

Sardinia has a unique and unforgettable scent, different from any other place I've known. It's the scent of the Mediterranean undergrowth, of junipers and myrtles. If I close my eyes and concentrate, I can almost smell it even now, thousands of miles away and many years later.

As a young girl, I anxiously waited for summer vacations in Sardinia where, every August, my family and I would spend a few weeks in a rustic oceanfront camp near the town of Posada. As I neared the end of the semester, still bent over my textbooks, I would imagine unending barefoot walks on floury sand beaches, hours passed floating in a tepid sea or exploring the shoreline.

Summer after summer, we return to the same place, shared only with a family from a small town nearby. We find them, as usual, curled up under a large umbrella that protects them from the intense sun. The parents are fully dressed in dark attire, typical for people their age in this region. The two daughters, about my age, wear lighter colors, but are still covered from head to foot. One knits a sweater; the other is absorbed in a comic book. Every morning, I pass by with my mom on our way to the beach. We say hello, as I parade by dressed in my tiny bikini that draws a cordial but disapproving look from the parents.

As we near the foot of Orvile hill covered by pines, a rider on a white horse leaves a cloud of dust while he guides his clanging cows. A flock of sheep takes shelter from the torrid sun in the shadow of an old cork tree, turning their heads as we walk by or scattering into the spiny underbrush.

My mother and I reach the dunes, where we set out our towels to spend the day baking under the sun, reading and swimming. Here I can swim and float for hours, until my hands and feet become so soft and numb I can barely feel them, or sit at the water's edge to spy on the hectic collection of creatures moving back and forth and digging holes in the wet sand.

The local shepherd Melchiorre hides behind a dune. He's certain we can't see him in his flattened position against the hot sand, but we know he's there. For a while, he stays still, watching us, then he disappears behind the dunes only to emerge again near the shoreline a hundred yards away. Stripping down to his underwear, Melchiorre plunges into the sea. My mom and I follow the powerful strokes of his arms until he vanishes on the horizon. We search for him carefully, waiting to see where he might resurface, but he never reappears. It's like the sea took him away to some remote shore. We are worried about him until the next day when we see Melchiorre, once again, lying behind the same dune, waiting for us. Day after day this scene repeats itself.

One morning, I get up early and leave for the beach alone. Melchiorre is already at his post, staring out from behind his dune. I lie down and open the first page of my novel. This time, perhaps because I am unaccompanied by my mother, he crawls through the sand like a giant spider, until he is standing next to me. He has dark skin consumed by the sun, and his hair is bristly and unkempt. His disproportionately short legs seem barely able to sustain the bulk of his muscular upper body. He may be in his thirties, perhaps even younger, but I can't tell. As he licks a couple of fingers to move a few locks of hair off his face, his broken teeth attempt a disorganized smile. He timidly talks to me in his strong, almost unintelligible dialect. He tells me that he lives in a small stone house behind Orvile's lagoon amid the rocks and shrubbery, with only his sheep for company. We spend the rest of that morning talking, and I tell him about where I come from.

Sometimes I see Melchiorre walking barefoot, guiding his flock toward the town of Posada. Local gossip is that he inherited a fortune from a relative and keeps the money hidden under his mattress. If that's true, the money certainly hasn't changed the way he lives. Melchiorre's days run at the slow rhythm of the tides. There is no doubt he is a strange fellow, as many people around here seem to think. He's so reserved and timid that some believe he might be a little slow. But to my young eyes, he's just a guy who doesn't fit into the modern world. Happiness, for Melchiorre, is his love of the open water and his long swims out toward the horizon. To me, he seems the embodiment of a simple life lived in contact with the sea. Over time, we become friends, and Melchiorre introduces me to a new world of abundant and diverse marine life, hidden among the rocks and tide pools of this undeveloped place.

The rocky shore in front of where my family is camped is still mostly untouched by civilization. The half-moon beach is covered by the brownish, salty hair of dead sea-grass leaves of *Posidonia oceanica*. Early in the morning, my father and brother leave to go spearfishing along the coast. Instead of going with them, I decide to stay near camp and play in the tide pools of the intertidal zone. It's enough for me to sit on a rock with my feet in the water and peek into this uncontaminated microcosm with its crowds of small and colorful sea stars, crabs, and tiny fish. I can easily lose myself here among this diversity of life I've found between the rocks and the sea.

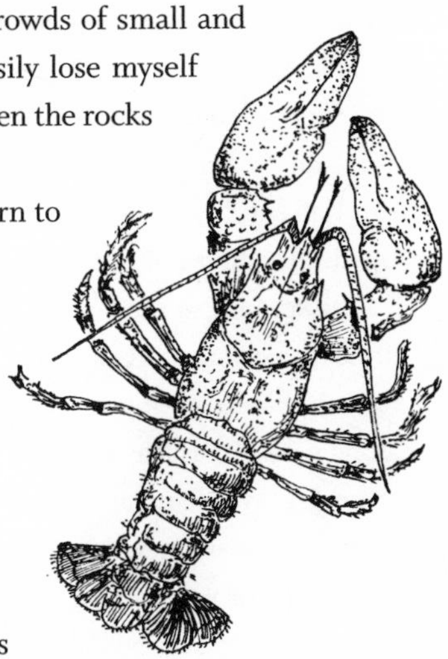

A few hours later, my father and brother return to camp. They have had a very productive day.

"Look what I got today!" my dad tells us, as he digs a lobster out of a net full of fish. "And you won't believe what Gio caught! It's huge . . ."

My brother proudly holds up a lobster the size of his arm. "It was hiding in a forest of *Posidonia*," he says. "I saw the antennae sticking out and grabbed them . . . just like this!" He gestures wildly with his free hand to illustrate his point. Orvile is teeming with lobsters. My father taught me early on that, even though the sea was rich with tasty fish, we should never take more from the ocean than we could eat.

I can't spearfish, but I am good at catching octopi. There are so many in Orvile that my dad and I simply grab them as they leave their hiding places in the rocks. On one occasion, an octopus twisted all eight tentacles around my arm, sticking its suction cups to my skin. I was trying to uncoil it, but the octopus was stubbornly glued to me and didn't want to let go. I finally gave up, and plunged my entire body in the water, hoping the octopus would loosen its grip. It worked, and my would-be prey quickly escaped. I was left with nothing but itchy skin covered with cherry-colored spots.

I had fun as a child and a teen; in those summers spent in Sardinia and along other shores of the Mediterranean, I explored the sea and nurtured myself from it. I recall lobsters and succulent oysters as being large and plentiful. I remember groupers and sea bass large enough to feed my entire family. But the world I experienced doesn't seem to be there anymore. The soothing sound of nature, the untouched land and open shorelines, the salty sea bursting with life are all disappearing. Large groupers and lobsters are gone. And the rest is vanishing with every day that passes.

As I travel back to the places of my youth, the same Mediterranean Sea that nourished its people for millennia is now a warmer, dirtier, and more barren place. We still know little of this underwater world that will likely collapse before we are able to learn its secrets. Not only are large top predators like sharks, albacore and bluefin tuna, marlin, and swordfish on the brink of extinction, but smaller species like anchovies are now becoming scarce. Overfishing is taking a heavy toll on fish stocks, and catch sizes are becoming smaller every day, yet we continue to fish further and further down the food chain. Soon, if we keep up this behavior, we'll be seeing invertebrates and jellyfish as the common fare on the dinner tables of the world.

But looking at the Mediterranean gives me only a glimpse of the dour future of Earth's oceans. Worldwide, unregulated commercial fishing fleets are pillaging our waters with ever-advancing technologies. Target species are plundered along with a massive bycatch (the unregulated, accidental capture of nontarget animals). We disregard and degrade the ecosystems in which these creatures flourish, slowly transforming them into dead zones and marine graveyards. And many of us never think about the consequences of these actions. It's not just about the survival of the species in question, it's about our own survival as well.

Today, nearly 90 percent of large predatory fish have disappeared. I sadly read that prominent scientists believe we face a collapse of entire fisheries by 2050. Some species, like tuna, may be gone in the next few years. Tuna, like the bluefin, are astonishing hunters. They are what marine biologist and writer

Richard Ellis calls "a quintessential ocean ranger, the wildest, fastest, most powerful fish in the sea."

I clearly remember Atlantic bluefin in the Mediterranean Sea as I grew up. I once watched them from the bow of a ship off the coast of France, as they cleverly herded a shoal of sardines. On a fishing boat off Tunisia, I observed a large school of bluefin pass under the boat at full speed on their way out toward open water. At sea, bluefin were common. They could be as large as a dolphin, and their massive muscular strength was evident as they swam to speeds in excess of twenty miles per hour. What I couldn't know back then was that bluefin tuna would turn out to be one of the most endangered species in the oceans. They now risk extinction in just a few years. And bluefin tuna are not alone in the race toward total annihilation . . .

My brother, Giovanni, for many years the president of Tethys, has been studying different species of dolphins in Croatia and Greece for over two decades. His research shows that the once-bountiful Mediterranean population of short-beaked common dolphins is now in a heartbreaking and dangerous decline. On the island of Kalamos, where Gio maintained one of his research stations, common dolphins have almost completely vanished from the surrounding coastal waters. Only a decade ago, things were different. Gio told me he used to relax in front of his research station and comfortably wait for the schools of common dolphins to pass by in the channel.

The sightings that were once an everyday occurrence had become, at best, a monthly encounter. As my brother's research efforts grew with the passing years, the number of dolphins dramatically plunged, from 150 animals recorded in 1995, to only 15 in 2007. At first, he thought the dolphins might simply have moved away from his study area, but further research proved him wrong. Most of the dolphins were missing.

My brother believes the main culprit in the disappearance is overfishing, primarily due to the recent arrival of large commercial purse seiners in the area, which have depleted the sardine and anchovy stocks. These species are what common dolphins eat. Gio explained to me how the traditional artisanal fishery, which sustained local villagers for centuries, didn't deplete the fish population. It was the development of industrial-scale fisheries that actually sucked life out of this place.

Over the years, Gio and I have swapped opinions, thoughts, and drafts of papers about the animals we are both passionate about. More than an ocean apart, we both conduct scientific research in the wild. We've lectured and written about cetaceans and the oceans and worked hard in our respective nonprofits to promote conservation. What for both of us started exclusively as research became, as we aged and the natural world around us changed, a continual effort to protect dolphins and their habitats. Neither of us deliberately chose this path, as at the outset of our respective careers, we could not have known it would be necessary. But the natural world changed, and we both changed with it.

It's a while now since Charlie and I left our home in California. We followed the Mexican border from California to Port Isabel, Texas, then turned left when we hit water. We traveled the US coast, continuing to head northeast toward Labrador. We are staying near a small fishing village next to Grand Jardin on the way to the rugged headland of Cape St. George, on the western side of Newfoundland. Charlie wants to kayak out to the end of the peninsula and explore the shoreline that is only accessible by sea. One of the locals warned him it may be dangerous to paddle there, especially leaving late in the day when the wind and currents are unusually strong, but that just makes the idea more exciting for my husband. Burbank and I decide to follow him on land, hiking along the steep bluffs that overlook the rocky shoreline. The scenery is amazing, and it's a great opportunity for me to spot minke whales that, for the last week, are moving back and forth less than one hundred yards from the coast on this stretch of sea swept by the cold Labrador Current.

As I help Charlie prep his kayak in the empty port, a curious crowd of men begins to gather around him. A few minutes ago the town seemed deserted, but now it's alive with people. They are all here to watch this tall, wet-suited man launch his bright yellow kayak from our car with California license plates. Not an everyday occurrence in this neck of the woods.

It's a regular workday in town, but few people are working. Nearly everyone in this place lives on government subsidies since the cod fishery crashed, some years ago. This village, like many coastal towns in Newfoundland, was once a place where cod were plentiful. But it isn't like that anymore.

"The sea there is full of fish that can be taken not only with nets but with fishing baskets," John Cabot's crew said, after its 1497 voyage to these same waters. But by the end of the 1960s, the cod catch peaked, and by the 1990s, the entire Newfoundland cod fishery had collapsed. What once was an unparalleled fishing ground that fed and sustained the population for centuries is now a desolate patch of water.

We leave Newfoundland for Burbank's "homeland" of Labrador and, aside from the occasional clouds of biting black flies, which don't leave a good impression on our dog, the countryside is truly rugged and amazing. The Trans-Labrador Highway is a gravel road that stretches more than three hundred miles across what seems an untouched and unspoiled wilderness. It is not unusual to encounter bears lumbering down the road or herds of caribou moving to new feeding grounds. But a closer look reveals that this remarkable place is home to some of Canada's largest hydroelectric developments, and extensive logging operations now threaten the once-plentiful caribou herds.

We head on toward Quebec City and the Gulf of St. Lawrence, where I hope to meet up with a colleague of mine from UCLA who is working on his PhD on blue whales. One good thing about being a marine biologist is that I have friends and colleagues all over the world, often in stunning and remote locations and natural reserves. It's a great way to see places and animals that might otherwise be off-limits to tourists; here in the Gulf of St. Lawrence, I was hoping to ride along and see the beluga whales common to this area. But this time, I'm told that my colleague is away visiting his girlfriend, so Charlie and I end up watching these fascinating white whales from shore in the pouring rain before heading back south.

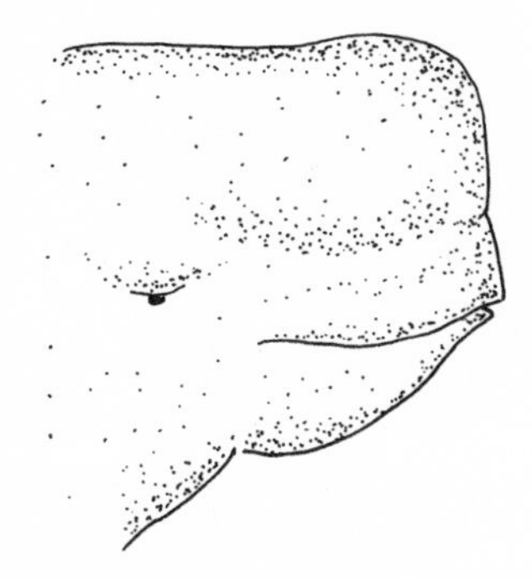

We settle for a few months near Annapolis, Maryland, where we rent a house in the woods on the shores of the Severn River, just a few miles from Chesapeake Bay. Edie Widder told us that she was interested in setting up a monitoring project in the Chesapeake, and we decide to stay and investigate the possibilities of incorporating a study on dolphins. Chesapeake Bay is home to one of the largest populations of blue crabs in the world, but in recent years, that population has been in decline. While the precise reasons for this are not yet clear, it may be a result of bacteria, viruses, and pollution originating in chicken farms and processing plants in Maryland and Delaware. Bottlenose dolphins are mostly found at the mouth of the bay, a hundred or so miles from where we are. In the process of working out the logistics for a preliminary study, something happens that drastically accelerates our plans.

Returning one afternoon from a day of kayaking, I find an interesting message in the OCS in-box. It seems that a lawyer from Louisiana wants to donate his boat to our nonprofit. We have had a few vessels donated in the past, but they are usually older boats that people are having a hard time selling. This time, however, the boat is a late-model Nordic Tug trawler that not only has a high value but is the perfect vessel for the kind of offshore research we conduct back in California. From the pictures included with the e-mail, *Annie Jo*, the boat in question, seems in perfect condition.

"It's too good to be true," we both mumble, but we are thrilled by the possibilities.

Charlie gets in touch with the owner, and we decide to meet. We have been traveling for some time now, and while it has been a fantastic and fulfilling experience, we have not been successful in setting up another project or collaboration with funding potential. I miss my research and being out at sea. This boat may be what we need to continue our dolphin studies in California. So our journey now takes a thousand-mile detour toward Port Kemah, Texas, near Galveston, where the *Annie Jo* is currently docked.

From the land of crab cakes, we head back to the heart of barbecue, where we meet Roland and his wife, Janet, the owners of *Annie Jo*. Roland has done some digging on OCS and is interested in the work of our nonprofit. He comes from southern Louisiana and is quite a character. I think both he and Charlie recognize a little of themselves in each other, and they seem to hit it off immediately. It turns out that they do, in fact, have a lot in common. They were both a bit wild when they were young, both sailors, both motorcyclists, and both seem to flourish outside of the norm. Before we know it, Roland and Janet have given the boat to OCS, and we are on our way back to Los Angeles to pick up research where we left off.

In less than a month, we are back with the large and comfortable trawler that we have shipped here from Texas. We find a new apartment and resume full operations with OCS. We have borrowed $100,000 against the future sale of *Annie Jo*, which will pay for our research for a while. My old team of LADP assistants is not only happy to be back at sea, but is pleasantly shocked by our new means of transportation. *Annie Jo* is much different from the boats we have used in the past. We've never had a research vessel where one can stay dry and warm while following dolphins, or sit comfortably on breaks, sipping a coffee or hot chocolate on the couch. We have never had a boat with a real bathroom, let alone a microwave, and hot and cold running water in the shower! *Annie Jo* is so luxurious that Charlie and I become almost paranoid about keeping her in pristine condition.

"I was going to clean that . . . but my God, who are you, Speedy Gonzales?" Shana teases me, as I follow her around, wiping up some milk that has spilled on the table.

Heading back home aboard *Annie Jo* after a two-day dolphin survey, we sight a gray whale mother and her newborn. It seems they are taking a quick digression inside the bay, before resuming their northward migration toward their feeding grounds in the cold waters of Alaska. A powerful blast of air is expelled from the mother's blowholes as she surfaces, and I can see the large patches of grayish-white barnacles encrusting her head. These sessile filter feeders "hitchhike" on the skin of whales and other slow-swimming marine organisms like sea turtles, over time forming colonies on the bodies of their hosts. These barnacles are not true parasites because they don't harm the whale or feed upon it. They are simply freeloaders along for the ride. They probably came aboard in the warm lagoons of Baja California, where this whale mother gave birth to her calf.

Some scientists use these barnacle patches to recognize individual whales because the patches are unique to each animal, just like the notches on dolphins' dorsal fins. I take pictures of the mother and calf pair, hoping I can match them with other images I already have of gray whales. They surface, this time with the calf swimming at its mother's side; then they are gone. They will travel day and night, covering over sixty miles per day, in what is one of the longest known annual migrations of a mammal.

But in the last few years, these migrations are getting shorter. Even if "full" migrations are not always the norm for gray whales, as climate change causes a decrease in their food supply of fat-rich shrimp at higher latitudes, some whales seem to end their journey in British Columbia, Oregon, even here, in California, scrounging for whatever food they can find. Like the dolphins of Kalamos, they too appear to be vanishing from areas where they once were common.

"Look how cute it is!" a girl dressed in pink tells her mom. She's watching a young twenty-foot gray whale floating near the end of the Marina del Rey jetty. Another woman in gym clothes is being interviewed by a television journalist who just arrived for a peek at our new local celebrity, the L.A. whale.

The small gray has been here for almost a week, drawing a crowd of residents and tourists snapping endless photos with their cameras and cell phones. They sit on rocks with binoculars, waiting for a glimpse of the "sociable" whale. The creature doesn't seem bothered by the crowd, and passes the time foraging on the shallow bottom at the mouth of the harbor. Occasionally it meanders out of the main channel into the open bay, but returns, shortly thereafter, to precisely the same spot near the jetty. The little whale eats, rests, and swims in circles, at times peeking out at the noisy swarm of people and small boats gathered around it. The whale is probably on its way back to Alaska, but day after day, as

I watch from my research boat, recording its behavior on video, it doesn't seem in any particular hurry. After ten days of being in the spotlight, the young whale resumes its migration. This kind of behavior is rather unusual for grays, and why it stopped here will remain a mystery. Perhaps it was separated from its mother, lagging behind a migratory group, or perhaps it was hungry and needed a break from the travel to recharge itself for the journey north.

Gray whales are adaptable, but not immune to human-induced environmental changes. Undernourished gray whales are spotted more frequently than in the past along the coast, especially in Baja California. Their thick and protective layer of blubber, the energy reservoir that helps them survive for months without eating, is getting so thin in some individuals that one can almost see their bones under the skin.

The problems are complex. We can't just blame overfishing for wiping out biodiversity and not allowing the oceans to replenish themselves. Overfishing is one of the biggest problems, but it's part of a larger series of issues. The real threat comes from the synergistic effects of human activities on the marine environment. Oceans are affected by climate change and sea levels are on the rise. There is more pollution, which opens the door to new diseases. Our oceans are becoming progressively more acid, because of increasing carbon dioxide in the water. Industrialized agriculture runoff has "fertilized" nearshore habitats, forming massive dead zones, where no creature can survive.

I read about environmental problems every day, and it is often overwhelming. I think I understand why many people feel paralyzed and unable to act. We are confronted with too much information to process. The news is often bad and frequently contradictory, so it is difficult for the public to know what is true and what is not. Adding to this confusion is the constant bombardment from television, e-mails, and Internet, which permeates all aspects of our everyday lives.

I am sad when I think that the world I grew up in has changed to such a degree that the encounters with nature I experienced may become just a memory. If we don't act together and soon, some of the magnificent creatures I've studied might not be around much longer. Time is *not* on our side. But I have a profound trust in our inherent ability to change and become better humans: more considerate of the world we live in and the diversity of life upon which we depend. This trust keeps me going.

Many years after the summers I spent vacationing in Orvile, I receive a letter from the shepherd Melchiorre, whom I have not seen or heard from since I was fourteen. How he found me is a mystery. In the last two decades, I've moved so often that I can barely keep track of myself. The envelope is handwritten, and my name is misspelled. I open it to find a two-page letter, written in pencil in a tentative and trembling script.

Melchiorre writes that he still remembers me as the young girl from the "continent," who came every summer to the beach at Orvile. He is writing this letter because I was the only friend he could talk to. He tells me he is, almost daily, overcome with a deep sadness. He describes it as a sadness that is choking him and for which he's needed to spend time in the local hospital. He writes about his long swims at sea, about how the only time he feels free is when he is far away from shore in the arms of the open ocean. He tells me of one day when he swam toward the horizon with only the thought of disappearing forever, but he was called back to land by a voice. A voice he thought might be that of God.

"It was a miracle," he writes, a vision that gave him the strength to return to land for another day on Earth. "I think I am OK now," he went on, "because I know that as long as I have my own place in nature, as long as I have the sea as my companion, I don't need to fit into the world of people that I have never understood."

As I fold the letter and tuck it back into its envelope, I feel a wave of warm, old memories sweep over me. I can see Melchiorre behind his dune, and I remember our talks and how he introduced me to his secret and special natural places, the places that are home to him. I can smell again the intense scents of Orvile, of the beach and the underbrush and the sea. And I will remember forever the peace that I felt in the natural purity of that place. And I cry.

PART 4

Learning from Dolphins

16

Alike

In the turbid salt marshes of South Carolina, as the tide ebbs, exposing a belt of mud, a flock of seabirds line up along the water's edge. They know it won't be long now. The creek is peaceful, and the still surface is ruffled only by the concentric rings left by the leap of an occasional small fish: a sign that they may be coming.

As a squad of Atlantic bottlenose dolphins approaches the shoreline, the birds noisily take to the air, hovering and ready to attack. The dolphins move in rank formation, creating a bow wave with their perfectly coordinated movements and driving a school of fish toward the muddy band of beach. Leaping to escape the dolphins, the fish land, stranded on the shore. Lunging after them, the dolphins push themselves onto the mud banks to snatch their stranded prey with quick and powerful body movements. The mud wrestling ends as the dolphins slide back into the water with mouths full of fish, disappearing beneath the cloudy surface and leaving the swirling flocks of birds brawling for the leftovers. Coming ashore with such violent determination, it almost seems an attempt to return to a long-lost terrestrial past, when the first dolphin ancestor dipped its hoofed foot into the sea over fifty million years ago.

Thousands of miles away, on the African continent, another coordinated attack is taking place. The assailants here are not dolphins, but chimpanzees.

As a band of adult male chimps strolls through the forest in search of ripe fruit, they come upon a small group of red colobus monkeys perched on a tree branch. Chimps and monkeys are aware of each other's presence, but the colobus don't move, as though they do not perceive any imminent threat from the chimps. A few minutes pass before the predators make a move. The monkeys panic, making screeching alarm calls, but it's too late. The chimps have mounted an organized assault, targeting colobus mothers with babies. The bloody conflict ends badly for the monkeys, as the chimps walk away, dragging their loot of dead monkey bodies. The meat will be either shared among the chimps or traded away as a sort of bargaining chip to negotiate sexual favors with females or establish dominance over other males.

"Hello, this is Craig Stanford," says the voice at the phone. "I work at USC and I just read an article in the *Times* about Dr. Bearzi's dolphin research. Is she around?"

"I'm Maddalena," I respond, " . . . Bearzi."

Craig introduces himself. He's a biological anthropologist at the University of Southern California and codirector of the Jane Goodall Research Center in Los Angeles. He has worked extensively with great apes and monkeys in Africa, Asia, and Central and South America, focusing attention on the ecological relationships among primates in tropical forests. He's also the author of several books.

I am familiar with some of his work, having read one of those books. I am excited to talk to him, as next to dolphins, great apes have always been the mammals that fascinated me most. Now this world-famous anthropologist was asking whether he might come out on the boat with me on a dolphin survey. We talk for a good half hour, and Craig seems far more interested in knowing about my work with dolphins than telling me about his own research on primates.

Next time I speak to Craig is aboard the LADP research boat. He is a tall, friendly person with dark hair, dark eyes, and dark skin. He's just returned from Uganda, where he spent a month with chimpanzees in the Bwindi Impenetrable National Park. He speaks fast and with much enthusiasm, as we talk about our interests and our respective research. As it turns out, Craig and I share not only a passion about dolphins and primates, but about all sorts of reptiles as well.

The hulky head of a northern elephant seal with its large proboscis, shaped like a short elephant's trunk, sinks into the sea ahead of us. A few minutes later, a school of Dall's porpoises glides near the surface. Dall's are remarkably fast swimmers, capable of speeds over thirty knots, but today they parallel our course at a pace we can follow.

"You're lucky," I tell Craig. "We don't see elephant seals or Dall's often . . ." In fact, it has been almost a year since my last sighting of these small porpoises, and in over a decade, I have counted no more than a dozen elephant seals.

After leaving the Dall's, we continue offshore for another hour or so before we sight a school of long-beaked common dolphins. The school has just begun feeding on a shoal of sardines that we can see swirling below the surface. Craig stands up on the starboard gunwale of the boat, watching intently as the dolphins work in unison to herd and catch their prey.

"You know, Maddalena . . ." he says, after watching the dolphins for a while, "it's interesting . . . the more we talk and the more I see, these dolphins, sort of remind me of chimps . . ."

As Craig and I continue to chat about our respective investigations on dolphins and great apes, something intriguing begins to emerge. At first glance, primates and cetaceans appear very different, but there are many striking similarities. From a behavioral perspective, these two taxa resemble each other more than other closely related species. Although dolphins and great apes live in opposite environments and haven't shared a common ancestor in almost one hundred million years, they are alike in many ways, more so than Craig or I ever imagined before our encounter. These animals are living examples of what scientists call convergent evolution: two completely diverse species with unrelated lineages that have evolved similar features.

Craig and I watch a couple of young individuals parallel the boat as they rub and touch each other. In the underwater world, these animals use contact as a way to reinforce social relationships inside the school, not unlike chimps, which use grooming to maintain bonds beneath the forest canopy. Craig and I are thinking the same thing, and by the time we get back to port, we have decided to look into writing a review paper on the comparative ecology of great apes and dolphins. What started as a day out on the water would turn into a long-term friendship between a primatologist and a cetologist.

There are a growing number of studies on the ecology of apes and dolphins, but in the course of writing our paper, neither of us can find any books comparing these animals' social structures and intelligence. So, after a few meetings, we decide to embark on a new adventure together and write one. Writing a book with Craig is a huge endeavor for me, and although I have written many journalistic articles, a book—in English—is another beast altogether. It forces me to push myself well outside my comfort zone, and I find myself having, once again, to wrestle with my old, familiar insecurities. Charlie helps throughout the entire writing venture, as a sort of "personal editor," and somehow I get through it.

A couple of years later, our book on great apes and dolphins is published. The book gives me a new perspective on dolphin research. Blending my new knowledge of great ape behavior with what I already know about dolphins affords me a better picture of their large brains at work in the open oceans. Elements including complex communication, social interactions, problem-solving skills, strong bonds between mothers and calves, care for young, and the capacity for emotion are illustrative of intelligence transferring into action. Not only is it now easier for me to see clear parallels between great apes and dolphins, but I begin to notice some of the same similarities between them and us as well.

"So, Dr. Bearzi, how smart are dolphins?" the newspaper journalist asks me, as we are leaving the harbor. Today this forty-something blond woman, dressed more for a Hollywood party than a ride on a research boat, is aboard to interview me about my work and about dolphin intelligence.

"It all depends on how you define and measure intelligence," I answer, adding, "by the way, please call me Maddalena." She smiles and begins scribbling notes in a pocket-sized pad.

Since the book came out, I've been asked to do several radio and television interviews, and I'm getting more accustomed to speaking about my work in front of a microphone or a camera. But I'm still nervous that I'll misunderstand something or not pick up on some nuance. English, after all, is not my native language. So far, it hasn't been a problem, so I suppose it's just the usual ghosts of my own doubts following me around.

I explain how one can make a case for intelligence in many animals, whether it's a dog or a cat or an octopus or a bee. One can look at animal intelligence in terms of problem-solving capacity or as a sum of attributes such as understanding abstract thoughts, language, reasoning and remembering, ability to use tools, self-awareness, and learning abilities. Some measure animal intelligence in terms of brain "contents" and encephalization quotient; the latter is defined as the ratio of an actual brain weight of an animal to the expected brain weight of a typical creature that size. There are many examples of "intelligent" animals, from primates to invertebrates, and many definitions of what "intelligent" might mean, depending on whom one talks to.

"But however you or I choose to describe intelligence," I say, "dolphins, and great apes for that matter, have many of the attributes we ascribe to ourselves." I go on to tell her that only in dolphins, great apes, and humans do we find brain complexity, social complexity, and ecological complexity so strictly linked.

I'm trying to make my point when I notice the journalist seems distracted. A large set of swells has just slapped *Annie Jo* on the port beam, making our deck roll like a rocking chair. The journalist is turning a distinct shade of green.

"Maybe we should take a break and you can have something salty, before you . . ." I don't have time to finish my sentence before she turns around and throws up over the railing.

"Sorry, Dr. Bearzi," she says, cleaning her mouth with her notepad and smearing lipstick all over her face. "My editor sent me aboard for this interview even though I told him I always get seasick . . . I even get carsick . . . Can you believe it?"

I give her some pretzels to help stabilize her stomach and one of our LADP foul-weather jackets to keep her warm.

"You'll feel better in a minute," I say, then I tell Charlie to head back to port.

Back on land, the journalist's natural skin color returns. She removes the jacket, adjusts her blouse, and fixes her makeup, and we pick up our interview where we left off.

"So, Dr. Bearzi, do you think we are close to apes and dolphins in terms of our intelligence?" she asks, scrolling through her notes.

"Really, call me Maddalena," I insist. "Well, the connection between great apes and humans is easy to grasp . . . after all, we share 99 percent of our DNA, and our brains have many similarities. We even look alike, perhaps more than we like to admit, give or take some excess body hair."

She smiles without raising her head. "And what about dolphins, *Mandolina*?"

"With dolphins, one needs to stretch the imagination a little further to see the link, but the connection is there. Dolphins have many of those attributes of intelligence I mentioned to you earlier. Their brain has similarities with ours, and their encephalization quotient is second only to humans." I tell her how, at sea, I've seen the plasticity of these animals when coping with their complex environment and how their brainpower is used to solve problems and develop social skills.

"Food resources in the oceans may change quickly and without warning," I tell her, "and as food is patchily distributed for both, great apes and dolphins, finding it requires flexibility and cooperation, both of which are indicative of intelligence. Cooperative hunting, in killer whales for example, is really intelligence in action." The journalist thanks me and waves good-bye as I head back to help my crew finish the last of the boat cleaning chores.

"So, Man-do-lina," Charlie says, as he picks up a pair of research cases, "lunchtime . . . ?"

I've always wanted to go to Patagonia. It is one of those far-off places that you never seem to get to. When I was still in school, I voraciously read the accounts of famous whale biologist and bioacoustician Roger Payne, known for discovering humpback whale songs as well as for his valiant and relentless efforts to curtail commercial whaling. For several years, he lived on the Valdés peninsula, in northern Patagonia, with his wife and children, in a place called Lote 39 (aka the Whale Camp), to study the little-known right whales of the Gulf of San José. In Peninsula Valdés, one can also find one of the most remarkable examples of cetacean cooperative hunting in the world.

On the beaches at Punta Norte, at the northernmost tip of Peninsula Valdés, killer whales stage coordinated attacks on seals that incorporate an advanced degree of adaptation, cooperation, even culture. It is adaptive because their technique has evolved over generations into a highly effective hunting strategy, unique to this place and to this prey. It is cooperative because the orca pod works together to achieve a result that is beneficial to the entire group, not just a single animal. And it is cultural because it is taught and learned and passed from adult to calf, generation after generation.

The pod approaches the beach silently, swimming underwater without using their biosonar, to maintain the element of surprise. The seals, unaware of what's coming, loll peacefully on the beach or undulate back and forth, enjoying the safety that land usually provides. As the orcas near the group of seals, they surge forward from the waves in unison, lunging onto the beach with their powerful bodies in synchronized perfection. The surprised seals, with nowhere to hide, are snatched left and right in the powerful jaws of the hungry whales. Twisting their bodies in a whipping motion, the killer whales retreat with their prey into the waves and the safety of deeper water. In teaching this technique to calves, adults must be careful that the calf, who follows the adult whale onto the beach, will be able to return safely to the sea. Failure likely means death. Once learned, the young whales will remember the maneuver for the rest of their lives, teaching it, in turn, to their offspring.

We are on a dirt road heading out to Punta Norte when a sudden sandstorm materializes out of a clear sky, slamming into the semi-functioning car we rented in the closest town, Puerto Madryn. Charlie slows the car, trying to stay on the road in the thick dust cloud that obliterates our visibility. A few minutes pass, and the storm abates enough for us to see again. Our car is now blanketed in a layer of golden-brown powder. I am not worried out here in the middle of nowhere—I am excited. As kind of a belated honeymoon, Charlie and I finally made it to Patagonia.

Peninsula Valdés is not like other places. Here one might see armadillos, foxes, maras, whales, penguins, guanacos, and elephant seals in a stretch less than thirty miles long. A volunteer at the local marine mammal observation station in Punta Norte tells me they spotted a pod of killer whales attacking seals just a couple of days ago. I explain to her, in Spanish, that I am a visiting cetologist and that I know some of the scientists working in the area. She invites me in and shows me a wall covered with the photo-identified dorsal fins of recently sighted whales. Then she directs us toward a good observation spot on a nearby hill where we can go to wait for the killer whales to come ashore.

"You never know . . ." the volunteer says to me, as I walk away, "they might come back today . . . maybe tomorrow . . . *Buena suerte, Magdalena*!"

We missed the season for observing now-endangered southern right whales (so called by early whalers because they were the right whales to hunt), which each year return to this uncontaminated stretch of Patagonian coast to breed. But even if now is not the best time to spot right whales, it's the perfect time for seeing orcas and the unique hunting behavior of these compelling top predators. With binoculars on our laps, Charlie and I sit at the designated spot and wait. As the afternoon wears on, I am captivated by the amazing scenery: the stark contrast of land and water against the dramatic backdrop of the unpredictable changes in weather.

Yesterday, we stopped at Caleta Valdés, nearly thirty miles south of here. We went there to see a small colony of Magellanic penguins sandwiched on a tongue of bushy land between road and sea. I had never seen penguins before, but I'd always been fascinated by these flightless migratory birds that glide efficiently through the water, using their rigid wings like paddles. On land, they saunter about like a funny bunch of miniature waiters.

It's breeding season at Caleta Valdés now, and the mated pairs nest in burrows. In these holes, the female will lay two eggs; then the couple will share all subsequent parental responsibilities. They will build their nest together, care for the eggs and future chicks, and take turns at home or hunting out at sea.

As we stood watching the penguins, a strong wind came up, driving sand and stinging pebbles through the openings in my wind jacket, which had inflated like a hot-air balloon. I climbed over the side of the bluff to get some shelter from the weather, and then I sat near one of the penguin burrows hidden beneath a shrub for nearly two hours watching an endearing pair of penguins. I couldn't take my eyes off them. They were gently touching each other, looking after their feathers and rubbing their heads together in what appeared to be moments of tenderness. It was a moving scene.

We are still waiting for the killer whales. The water is getting rougher as the afternoon breeze picks up, and hordes of whitecaps are making it almost impossible to spot whales. I turn around to see a gray Patagonian fox cutting in front of a young girl walking toward the observation station. The fox stops for a passing glance at the girl, then moves unhurriedly away in the direction of the underbrush. It does not seem afraid of the girl at all.

We have waited for many hours, but no orcas have come. As I sit, I play back the awe-inspiring scenes of the orca hunt I'd watched in so many nature documentaries. But nature doesn't work on command, and that day as well as the next, I miss my chance to see the synchronized attack of the killer whales. In consolation, though, I did spot three orcas with their tall fins slicing through the waves near shore, as if they were readying for an assault. But no assault ever came.

As we leave Punta Norte for the last time, we come across an Australian tourist in the parking area. He is bent over and running backward, holding a piece of bread in one hand and a video camera in the other. He's videotaping an armadillo chasing the bread around the parking lot. We watch this odd scene until the armadillo tires of the chase and runs back into its hole.

"Where did you guys come from?" he asks, surprised to see us.

"Over there," I gesture. "We were waiting for killer whales."

"Well, you should check out what I got on video," he says, grinning widely.

We watch the screen that's occupied with the nodding head of this prehistoric-looking animal in its protective armor as it trots quickly after the

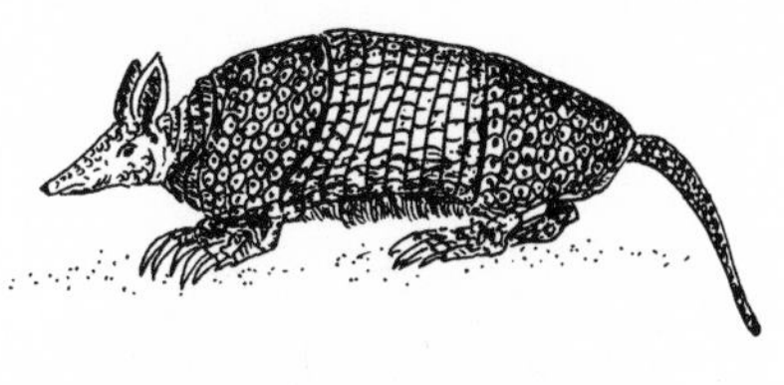

bread in the foreground, to the sound of the howling laughter of the Australian. I try to imagine what the armadillo saw, as it chased this screaming flabby biped backward around the lot.

"Jeez, he almost bit me, isn't that hilarious?" the Australian says. "Just wait, you're gonna see this on that funniest animal video show, you'll see . . ." He walks away in search of the vanished armadillo, hoping for more footage.

It was kind of funny, but it was sad at the same time. I, too, am a sightseer and certainly no better than anyone else, but I sometimes wish that tourism had never found these corners of unspoiled paradise like Punta Norte or El Palmar.

The diverse cooperative hunting strategies of dolphins and great apes are a good illustration of their social complexity and plasticity, but the strong mother-to-baby bonds and the teaching processes evident in how these animals raise their offspring are what remind me most of humans. Dolphin mothers and their offspring form a liaison that can last well over two years; in chimps these bonds can extend for a decade. During this formative time together, the offspring learn how to survive in a dynamic environment, whether in the ocean or in the forest.

From research boats in the Mediterranean and Caribbean Seas, in the Gulf of Mexico, and along the Pacific coast, I've seen how mothers patiently tutor their calves to catch their first fishy meals. I've watched mothers repeating the same actions countless times, teaching new survival skills to their offspring. I've observed calves carefully imitating their mothers' behavior and instruction. I've followed calves safely swimming in their mothers' slip stream, the wake mothers create as they move. Less than a flipper away, dolphin babies position their bodies next to the mother so the water she displaces while swimming reduces the drag and, consequently, the offspring's effort in keeping pace with its parent. A calf might abandon this comfortable position to explore the newness of the surrounding world with a child's curiosity. But its mother is always watching, supervising from close by.

It's true that dolphins and humans don't look alike. Dolphins have no hands with which they can eat their meals, no hair on their bodies, no vocal cords. We live in opposite environments. But some species of dolphins share remarkable similarities with both great apes and humans. Their ability to learn and transfer information, to communicate, to remember and imitate, to develop strong bonds and care for their young, and even to use tools and recognize themselves are all testimony to the common, intelligent traits that our species share.

17

Emotional Beings

Many years have passed since I first sighted Superhero from the beach at El Palmar. At that time, I knew little about dolphins and how they spend their lives. My naïve curiosity was initially driven by an insatiable thirst for knowing more about all things oceanic. I was intrigued by the idea that these sea-dwelling animals were mammals like me, even though they seemed so different at first glance. Now, after recording the complex societies of dolphins, I have formed a deep loyalty to these creatures. They are welcome companions in the presence of whom I always find joy.

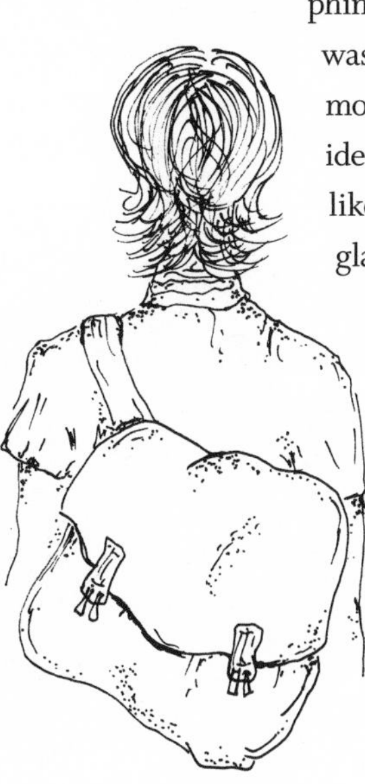

Over time, I've learned how they cleverly deal with an ever-challenging ocean; I've seen how alike they are to other, seemingly unrelated species, including my own. The more I have watched dolphins in the wild, the less I saw them as objects of my scientific research or as an anonymous and undistinguished group. Rather, I began seeing them as single individuals, not solely for their scars and notches, but also for their personality and emotions.

As fascinated as I was in those first encounters in Mexico, dolphins appeared to me then as identical creatures, lacking the expressive facial traits we humans depend upon to recognize the intentions or feelings of other beings; but they were not. They were not like the snakes and lizards I had studied before. I felt there was something more complex behind their fixed smiles and within their streamlined bodies. For years, I thought that feeling came from my anthropomorphizing, projecting my human values on dolphins to construct some interspecies connection out of nothing. I often reminded myself that I needed to remain impartial, critical, and systematic. On some occasions, I would let my objective guard down and go along with the moment, as I did on *De Bolina* and *Gemini* and later aboard my own research boats. But generally I tried to be detached.

At the inception of my studies on these animals, I was convinced that attempting to interpret some of their behavior in terms of feelings and emotions might cloud my ability to understand them in their own environment. Being detached is how I was programmed to think as a scientist, even if I've always been captivated by the magic of a dolphin encounter. Twenty-plus years ago, the idea that dolphins share similar emotions to those of our own species seemed far-fetched.

With more experience and time behind me, I began to better understand that the ever-present smile of a dolphin might hide a true smile or annoyance or grief—emotions not unlike those that I might feel. I also realized that I would need to learn how dolphins express those emotions in their own terms, making sure that I did not color those observations with my own human standards. Looking at these animals in the field and chronicling the nuances of their lives, I now think they are cognitive species capable of feeling emotions in their own way. I also believe that by opening our minds to better understand their emotions, we might learn something about our own.

Many animals experience emotions. I grew up with all sorts of creatures, and dogs have always been my closest friends and dearest companions, in moments of both joy and sorrow. Burbank, probably the best dog I ever had, often let me know how he felt. He showed me affection in my moments of sadness or sickness, by lying next to me and pushing his head and body against mine, as though he knew something wasn't right. And I recognized a sort of silent suffering, when he was ill or injured.

When Burbank was a puppy, I was throwing rocks along a dirt road in a desert canyon where we'd gone hiking. My dog, with the keen instinct of a retriever, was running back and forth along the road, finding the rocks I'd tossed. On one throw, I miscalculated the distance and landed a rock squarely on Burbank's nose. He looked at me with what seemed a sense of shock and wonder, as though he was thinking, "Why did you do that to me?" But he didn't react as if threatened. I immediately ran toward him, and it was clear that I felt horrible for what I'd done. A moment later, he was licking my face and wagging his tail as though nothing had happened. Somehow he understood I had made a mistake. Call it what you will, on that day I saw my dog express joy, surprise, confusion, and trust. Anyone who has ever owned a dog would have a hard time denying that these animals have emotions.

I have always been fascinated by the compelling accounts of animal emotions, and in recent years these accounts are becoming more common. Behavioral ecologists, ethologists, animal psychologists, and others are publishing new case studies, almost daily, about a wide variety of species that include lions, pandas, chimps, wallabies, foxes, mongooses, meerkats, camels, dogs (of course), cats, giraffes, elephants, horses, hyenas, even octopi.

Renowned primatologist Jane Goodall has devoted over forty years of her life to the chimpanzees of the Gombe Stream National Park in Tanzania. Her work paints a portrait of these animals' unique personalities. She has shed light on their capacity for love, altruism, cooperation, and compassion, as well as brutality and conflict. She has emphasized the parallels between them and us, and offered us a new view on the significance of being human. As her studies progressed, Goodall stopped calling her chimps "it." They became "she" and "he," and she gave them names like Flo and Fifi, a practice that raised eyebrows in the conservative scientific community of that time. In years spent among her forest friends, Goodall documented sorrow and joy, fear and physical suffering. She saw them kiss and hug each other, fight with anger, and pat each other on the back as old friends do.

Historically, cetacean research has lagged years behind that of primates, likely because it's easier to study animals on land than in the water. In my talks with Craig Stanford, he told me how he would sit for hours in the shadow of a tree, watching a group of chimpanzees and taking notes on their lives, which were unfolding in front of him. Dolphins, though, are hard to see and follow. They don't have arms to hug or the capacity to form facial expressions like their terrestrial counterparts. Nonetheless, I've been a quiet witness of some astonishing facets of their emotional lives.

I perch on the bow of the LADP research boat in the open ocean, watching five lively young dolphins fool around with a string of seaweed. One holds the kelp in its beak and teases its companions with passing glances. Then it quickly retreats with erratic tail strokes. Another dolphin attempts to grab the seaweed from its peer, but misses. A few minutes later, the first player releases the kelp, and a second individual grabs it. Now it's this dolphin teasing the others, while the four companions race after it in a three-dimensional, gregarious water game. I see social play; I see dolphins having fun; I see joy and curiosity. And it's like watching a group of kids playing football in a city playground.

Out at sea, I frequently observe empathy in these animals. I watch dolphins "caressing" each other with their pectoral fins in tender moments, as reciprocal body signals of affection. On those days when the water is so glassy I can easily play detective beneath that blue mirror, I might see dolphin couples engaging in their courtships. As a prelude to mating, they perform a sort of dance together. Belly to belly, they face each other, seeming almost as though each is the reflection of the other. They twist and nuzzle, lightly bump heads or gently tooth-scratch one another. To me, these complex interactions appear more than simply reproductive or instinctual.

I have witnessed as well the depth of a mother's parental love as she tirelessly aids her calf in taking its first breaths of air. In their company, I saw caring and compassion and emotional attachment. It was in Belize, toward the beginning of my days with dolphins, where I got my first look at how a dolphin mother cares for her calf.

In the turquoise Caribbean waters off an atoll tangled in mangroves, I leaned off the bow of a fisherman's pale reddish boat, observing a bottlenose mother and her newborn. As we approached them, I didn't take too much notice of the drama that was unfolding in front of me. I was still distracted by the accident of the previous day.

I had just arrived in Belize and settled into a cheap and somewhat shoddy hotel room complete with cockroaches in the shower drain and screaming kids in the hallway. The following day, I planned to meet with a local sea turtle researcher and discuss a potential conservation project down the coast. Then I hoped to go out and scout around for dolphins. Belize is another of those places in desperate need of conservation. Green, hawksbill, and loggerhead turtles use stretches of this coast and barrier reef as nesting and feeding grounds. But here, as in Yucatán, people still harvest turtle eggs for consumption and sale, and these reptiles are often accidentally killed in gill nets, long lines, and shrimp trawls.

"We have a curfew after 8:00 p.m. in Belize City . . . be careful!" the bus driver told me before dropping me off. I had traveled all over Central America by myself, and I wasn't too worried by his words. I assumed the driver was just exaggerating, seeing a young girl traveling alone. And I was hungry, so after checking into my hotel, I went out in search of a restaurant.

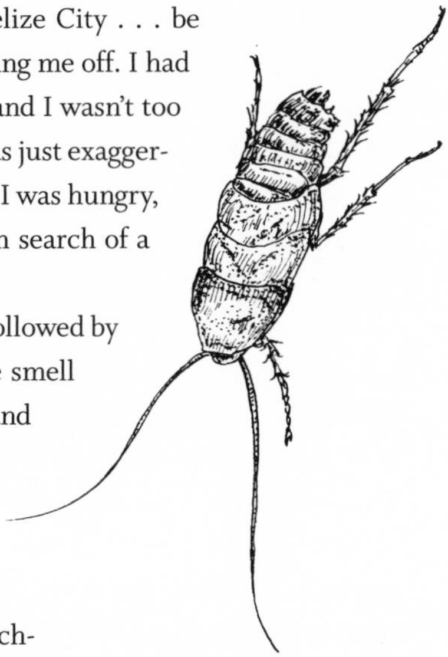

I walked along the shoreline promenade, followed by a mob of barking street dogs and the intense smell of lifeless seaweed. The evening was muggy, and I was wearing shorts and a tank top and an old gold necklace that my grandma gave me when I was a child. I followed the sign for a Thai restaurant down a side road when I heard the crackling sound of a bicycle approaching from behind.

I turned just in time to see a skinny, bearded man pedaling swiftly toward me on a rusty bike. In a flash, the man grabbed for my thin gold chain. Without thinking, I raised my hand to protect my necklace, and he lost his equilibrium. Lurching forward, he made another grab, and his long fingernails raked my skin, leaving four deep gashes in my chest. In the commotion, he jumped back on his bicycle and took off. Miraculously, he'd dropped my grandma's broken chain on the street in front of me. With blood all over my torso and tank top, I rushed back to my hotel where, in the company of cockroaches, I inspected my wounds. I looked like I'd been slashed by some animal's claw. I had always traveled alone, often in remote places that my friends thought were unsafe. I had never been robbed or hurt. Perhaps I was just naïve or immensely lucky or stupid, or some combination of the three. But I was safe and OK, except for the scars that would linger for years to come.

My attention snapped back to the mother and calf as, out of the clear blue water, I saw the calf struggling to lift its head at the surface to breathe. It seemed that it had serious problems breathing. The mother was relentlessly pushing her infant's body upward, so its small head could emerge, and its blowhole could open toward the air. But the calf wasn't learning the rhythmic motion necessary for breathing. For minutes at a time, I didn't see the pair resurface, and I was sure that the calf hadn't survived. But then I'd see them again, next to each other, as the mother continued her attempts to guide her offspring through its awkward and difficult struggle for air. I watched for over an hour as the newborn calf slowly began to breathe on its own. As we left the site, I could still see the mother carefully directing the newborn's actions, and I knew that the mother would continue to help her calf, no matter what.

There are times when the bond between a dolphin mother and her newborn is so strong that even death doesn't fully break it. With unspoken suffering and unshed tears, dolphins (and not only female dolphins) might mourn as we do for the loss of another human being. A dolphin mother may remain close to the floating, lifeless body of her dead newborn for hours, sometimes days, constantly trying to lift her offspring toward the surface with repetitive and anxious movements. She will often touch the lifeless body with her flippers and rostrum in a last hope to revive and breathe new life into it. But she never abandons her calf, not even to replenish herself with food. Her devotion is unselfish and unfaltering. A grieving dolphin mother may seek seclusion, away from her group, but in this time of grief, she might be visited by a group of her peers, perhaps coming to check on her, as we humans often do when someone we know is bereaved.

Just as affection and compassion can play a part in a dolphin's life, so too aggression and fear play a similar role in the oceans as they do with us on land, and I see examples of these emotions in my studies in California.

I'm catching up to a mixed group of bottlenose and common dolphins forty miles off the coast of California. There is a rainsquall on the horizon, and the water has become a bright tropical blue. The surface is broken every so often by the flashing bodies of flying fish. They move their tails at an impressive speed of up to seventy times per second to launch themselves from the surface. Then, spreading their large pectoral fins like a bird's wings, they soar through the air, staying aloft for distances that can exceed fifty meters.

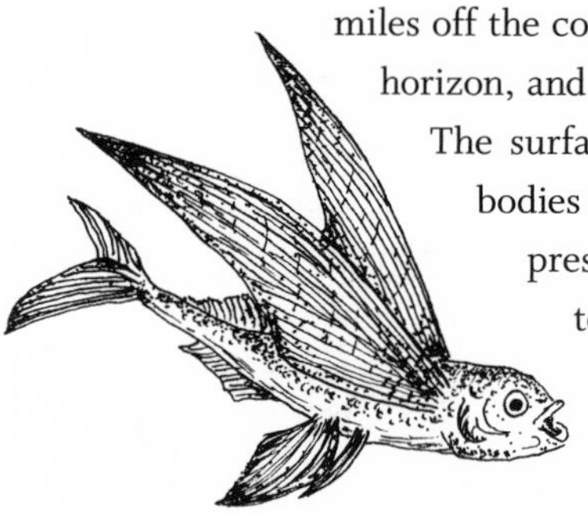

Hydrophones on, I hear a loud series of jaw claps: a clear sign of belligerence in the dolphin world. The clapping goes on for a few moments, and I can see a trio of bottlenose quickly circling a lone common dolphin that, lagging behind its own group, now finds himself confronting these three larger animals. The bottlenose twist and turn around the common dolphin, which is visibly trying to escape, like a frightened person would do if faced by a group of bullies on the street. Through all the movement and bubbles in the water, I can barely discern what's happening, but it's clear the situation is escalating into some kind of physical violence.

Through the hydrophones, we hear a cacophony of clicks and whistles, and I see one of the bottlenose in the gang aggressively lunge at the common dolphin, biting its flank. Ten or more common dolphins now arrive and surround the injured individual while the aggressive bottlenose promptly retreats with its companions. The common dolphin is now back in the safety of its own group, and they move away in the opposite direction from the bottlenose. The rain has reached us, tapping incessantly on our decks, and my team and I escape into the cabin and make our own retreat back toward land.

I always try to respect dolphins' "personal space" by maintaining a safe distance from them, but in many years at sea, the presence of my research boat has provoked some reactions of annoyance. I remember one male dolphin losing his temper in the twinkling of an eye. He went from peacefully riding our bow waves and glancing at me, to furiously slamming against the boat. I was attempting to shoot pictures from the bow, when the dolphin started striking the hull with powerful strokes. As I turned toward Shana to tell her to move the boat away, I was soaked completely by one of his violent tail slaps, camera and all. To this day, I have no idea what set him off.

Quantifying these emotions is another thing altogether, especially in the wild, where brain responses cannot be monitored. Knowing whether an animal is experiencing caring, playfulness, anger, grief, or whatever may be something that the observer only senses, something that, as yet, is difficult to state with scientific surety. Our own psychology colors how we interpret what we see. But many of us who spend our professional lives in the company of animals have the similar sense that there is feeling there, as well as instinct. There are times when it is hard to remain objective because we are affected not only by what we think are animals' emotions, but by the emotions stirred within us by the observation itself. One such instance that I will always remember occurred when a baby seal attempted to climb into the LADP boat.

I was at the helm of our old Bayliner, quite a distance offshore, when I spotted the tiny head of a harbor seal pup. I slowed the boat and approached warily, trying not to scare the seal. The pup did not seem disturbed by our presence and stared intently at us as our boat drew closer. Brigitte and Karyn carefully scanned the horizon for any sign of the mother, but there wasn't a trace of other seals in the area. This patch of ocean seemed deserted.

I was maintaining a safe distance when the pup emitted a barely audible prolonged squeaking sound and moved directly toward us. I shut off the engine and watched, as the little creature slowly swam up to our stern. As it stopped for a moment to inspect us, I could clearly see its sleek spotted coat. The pup's liquid eyes and long whiskers were so disproportionately large in comparison to the size of its head, it looked almost comical.

"Oh, she's sooo cute," coos Karyn, ascribing a gender to the baby seal. Not much of a scientific observation on Karyn's part, but she was right, the pup was irresistibly cute.

Then something out of the ordinary happened. The baby seal swam up against the side of our boat and, with swift movements of its flippers, attempted to come aboard.

"Oh my God!" exclaimed Karyn. "She wants to come with us . . ."

"Come on, Maddalena . . ." Brigitte joined in. "Can we take her?"

"Let's see what she does," I replied.

"We can take her to the Marine Mammal Care Center," added Paul, who had been videotaping the whole thing.

"Let's see," I repeated.

We all watched as this seal puppy tried again and again to get out of its medium and into ours, pausing every so often to regain strength. Every time we thought it would leave, it fervently scratched its small flippers against the boat, attempting to push its head as far out of the water as possible, as if to get a glance of what was inside our boat or catapult itself aboard. The pup was so repetitive and insistent that a wave of compassion overcame any scientific objectivity that I had attempted to maintain. Was this a desperate call for help? Was the little pup abandoned at sea and in search of any company it could find, regardless of whether it came from an utterly different species? Was it fear that this little animal was feeling?

My team were all looking at me with the same heartbreaking puppy-dog eyes of the seal, but I resisted compassionate involvement, deciding it was best not to interfere with nature. The baby seal seemed in good health, and its companions might have been close at hand. We were in their world, not ours. Before I had a chance to voice my decision and incite mutiny in my crew, the little seal gave us a last look and swam deliberately away.

That day at sea was a good example of emotion playing havoc with objectivity. The emotions of my crew, perhaps even those of the pup, were all intertwined together. All of us, including me, thought we witnessed a scared and lonely creature appealing to us for help. But none of us will ever know the reality of that situation or what the future held for that baby seal.

The more I learn about my fellow animals, the more I feel the need to protect them in their natural state. Many people ask me whether I swim with dolphins in my research or if OCS runs swim-with-the-dolphins programs. And I see their disappointment when I tell them that I do not, nor do I think that swimming with dolphins is a good idea.

I have never swum with dolphins, and we've never had such programs at OCS. Although I imagine that swimming with dolphins would probably be an amazing experience, in my heart, I know it is the wrong thing to do. As a cetologist, I would be setting a bad example. I have an enormous respect for dolphins and the oceans they live in. I feel privileged to have shared countless hours with them in the open sea. But I have never lost sight of the fact that these are wild animals, who need their own breathing and living space, just as much as I need mine.

I am driving down to San Ignacio Lagoon in Baja California to see the breeding grounds of the gray whales and visit the ecotourism programs going on there. Unlike many of the tourists coming to this area, who come to see the breeding behavior of these whales, I have come to observe whether I can identify any behavioral effects from the burgeoning ecotourism business here.

After driving over thirty-five miles through the salt flats and salt refineries that surround the area, I reach the lagoon, and, sadly, the sight is far from what I had hoped for. Local fishermen, turned guides, maneuver their *pangas* overflowing with visitors as near as possible to gray whale mothers and their calves. At times, the fishermen wedge their boats in between the two, to better afford the paying tourists a personal encounter with these gentle giants. It seems that everyone wants a photo of themselves petting, hugging, or kissing the head of one of these barnacle-encrusted behemoths. I can hear the excited screams of the tourists even from where I stand on the shoreline.

The ecotourism brochure says the whales here are "friendly" and come close to the boats for that reason. Most certainly they are curious. But being curious does not mean that all this activity does not disturb them. Nobody really knows what the lasting effects this continual, noisy stalking and touching might have on either mother or calf during that extremely fragile and formative period of their life together.

One person touching a whale, or swimming with it, is not the problem. The real issue is the hordes of people disturbing these animals with little control or regulation. I realize the benefits of situations where the public can develop some direct connection with whales and dolphins. It helps people learn the importance of caring for and protecting these animals. But whale watching must be regulated and conducted with the animals' well-being in mind, even if it means resisting the temptation to pat their heads. Now more than ever, as our human population nears seven billion, we need to preserve the wild spaces that remain on our Earth. This can happen only if we begin taking personal responsibility for our actions.

I read about the rising number of swim-with-the-dolphin programs worldwide. There are folks who believe that dolphins (and other animals) are there for their amusement. They believe that dolphins and whales "like" to be around us as much as we like to be around them. But that isn't always true. Dolphins might be inquisitive about our species, even sociable at times; however, like all wild animals, their behavior can be unpredictable, even hostile toward us. When we read of dolphins attacking and injuring humans, we are shocked that such a thing might happen.

We think, "How dare they do that to us?"

Recently the news of a killer whale attack on a human at SeaWorld in Florida captured the attention of national news media. A female trainer was slain by an orca, evidently without provocation or discernible reason. The trainer and orca had worked together for years until that dreadful day when something went wrong. When I heard about this, I felt sadness for the loss of a human life, but I also felt anger for the lack of consideration and understanding of these wild creatures that could bring anyone to keep an animal like that in captivity.

At the time I wrote a letter to a local newspaper, saying, "Anybody who has ever seen these animals at sea knows these are astonishing top predators, ranging over hundreds of ocean miles. . . . They display complex behaviors and have outstanding teaching and learning abilities. . . . Such an assault on a trainer should not surprise us. Wild animals are prone to act as wild animals even when caged. They don't perform to our species standard, and one should value them for who they really are. If we don't learn to respect these creatures by leaving them where they belong, and make efforts to protect their environment, there won't be any of them to see in the future."

There are huge economic interests behind not recognizing emotion in animals. If they have no feelings, it's easier for us to justify caging and using them for our own purposes, be it scientific testing, food, or entertainment. Well over a century ago, Charles Darwin wrote that the difference between emotions in humans and other animals may be just a question of degree. The wall dividing us and them as emotional and non-emotional creatures seems to be crumbling somewhat in recent years. We are beginning to embrace the controversial concept that at least some species might be not only cognitive, but also capable of emotions like joy, anger, or even love. Recent discoveries in the field of neuroscience show the brains of dolphins have mirror and spindle neurons. In great ape and human brains, these neurons are understood to play a key role in cognitive emotions, social cognition, and perhaps even the highly debated theory of mind: the ability of an individual to sense what another individual is thinking.

I remember buying my first camera and, with great excitement, walking into the zoo in Padua to photograph those great, wild creatures. But as I wandered past the cages, what I recall most were the heartbreaking expressions of broken animals locked forever behind bars. Now I know too well how dolphins are stripped from everything they need in a tank. Like any prisoner, a caged dolphin is not a joyful dolphin.

I'm the first to admit that emotion in animals is difficult to define and even more difficult to measure. This is further complicated by the fact that different disciplines mean different things when they talk of emotion. Just think about how hard it is for us to know what we ourselves are feeling in any given moment. But with my own eyes I've seen the societies in which dolphins live and how they find companionship and build lasting bonds, just as we do. They thrive in open space where they can move freely, communicate with others, and learn the culture of their species. The eternal smiles on the faces of dolphins in a tank should not be taken as an indication of happiness. The signs of their true emotions are there, if we can only be sensitive enough to notice them. Dolphins in captivity are nothing like the dolphins I've seen in the ocean wilderness.

I feel connected to the animal world by my humanity, not superior to it. My experiences have brought me face-to-face with animals in their own environments, where I have encountered diverse personalities and witnessed an array of emotional responses. As a result, I feel I am an integral part of nature, interconnected in the magnificent web of life.

Epilogue

The End (or the Beginning)

In wildness is the preservation of the world.

HENRY DAVID THOREAU

I'm swimming slowly and alone through crystalline waters beneath a cobalt-blue sky. My skin feels pleasurably numb immersed in this medium. I am totally free, as I turn and dive effortlessly toward the white sandy bottom. I can move as I choose, up and down with the motion of the waves in continuous harmony. I have no destination or future here, no resistance to this fluid world. I dream of being a dolphin without laws, clocks, duties, or clothes; liberated of earthly restraints and inhibitions; moving smoothly through the water with my powerful streamlined body in this three-dimensional life.

All around me, I see other dolphins. I feel I am part of a family, but it's somehow different from my terrestrial one. We are a grand and flexible family, yet I have a strong sense that our underlying bonds remain constant. I reach a huddle of mothers and calves. Seven females are clustered, forming a protective screen of bodies around a pregnant female. They swim beside and below her, waiting for the moment to arrive. In a cloud of blood, a newborn comes to life. Mother and offspring existences will be united for years to come. I watch as the new mother, her newborn, and all dolphins move away and disappear. I rush to follow them, but I can't. My movements become clumsy and my body heavy. I am unable to hold my breath any longer.

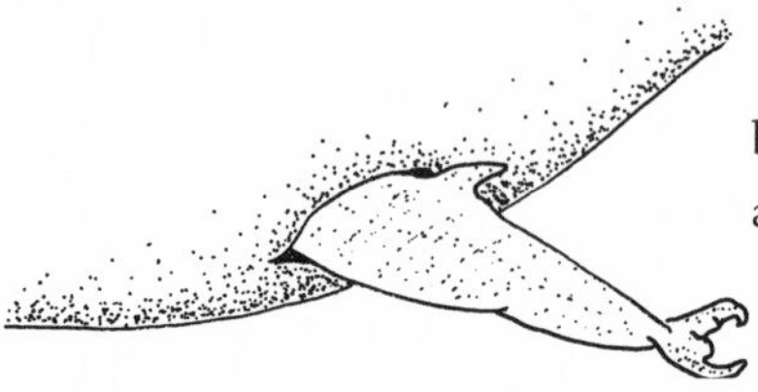

The phone rings. I fight hard to remain asleep, to stay with the dolphins. But it rings again, relentlessly. It's the end of my dream and the beginning of a new day on Earth.

I've been in the company of dolphins for more than twenty years, observing their social behavior, their emotions, and the striking similarities to our species. I feel at home with them. They are in my life, sometimes even in my dreams.

This intimate life with dolphins, whales, and other creatures has left a profound and beautiful impression on me. Because of them, I often question myself about my place on this planet, and what traces I will leave behind me. I've encountered many animals in my work, some more clever than others, but they all—in their own way—have taught me lessons of humility and made me wonder about humanity's conceit and insatiable greed.

When I first walked into the unknown natural world of animals, I had little skill or confidence. I was driven only by a great passion and curiosity for all kinds of living beings, with which I share tenancy on Earth. Guided by raw enthusiasm, I experienced wildlife firsthand as a child, venturing into out-of-the-way territory and meeting new animal friends along my way, both on land and at sea. Gradually, I became a woman and a marine biologist. A little at a time, I grew more at ease with my chosen vocation and more proficient in my work in the field. With age and experience on my side, some of my early insecurities, as well as the perceived boundaries left over from my upbringing, began to soften and dissolve, even if sometimes I still doubt myself, the significance of my work, and the meaning of my existence. My life, so far, has been an amazing journey between mind and nature, at times hard and unkind, but by and large rewarding.

I still feel that I know little of the complex lives of dolphins. Despite the years I've spent with them in the wild, I am struck by how much there is still to learn and discover of their ways and of their world. And at the same time, I grow increasingly worried about their future status and well-being. In the relatively short time I've devoted to studying dolphins, I've witnessed a dramatic increase in the number and magnitude of insidious threats facing these animals and their ecosystems. It has only been in the last half of the last century that we have realized the adverse effects of commercial whaling, chemical contamination, acoustic pollution, overfishing, bycatch, and human disturbance. And then there is climate change. For the first time in history, dolphin populations and even species are disappearing due to human impacts.

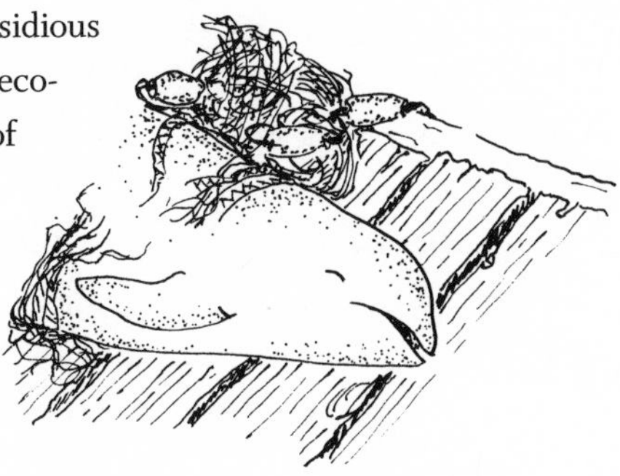

Nature will pull through any human-induced devastation and absorb the combined damage from everything we can muster against it. Some new balance will occur in which we will likely not be present. This may happen regardless of any human blip in our planetary history. But today we are the dominant species on Earth, with the capacity to change our ways. And it would be a sad and disappointing legacy if we didn't reverse the frightening upsurge of anthropogenic issues that are pushing nature, as we know it, to the limit.

The natural world of my childhood, the world where I first met dolphins, is not the same world that I live in today. Maybe it's always the case with each passing generation that the world we are born into necessarily changes as we grow older, but now the natural spaces are not just shrinking—they are disappearing. And I wonder how far back we'll need to turn the clock to even remember what was here before we developed it.

Real changes must happen now to protect the collective future of dolphins, oceans, and humans on this planet. But what changes? I have attempted to make a difference on any level, local or otherwise. And it isn't easy. Sometimes my efforts are successful, but many fail. Some give me hope or serve as the basis for new ideas, while others force me to reflect on what I did wrong in order to find some new path to try.

A holistic approach may go a long way toward solving environmental issues. In his book *A Sand County Almanac*, the great naturalist Aldo Leopold talked about a "land ethic." He described how "a land ethic changes the role of *Homo sapiens* from conqueror of the land-community to plain member and citizen of it. It implies respect for his fellow-members, and also respect for the community as such." Leopold wrote about land, but his words apply to the oceans as well.

Hope, love, humility, and compassion are key components that we often fail to remember when it comes to making our natural world a better place. We are so consumed with our everyday lives, there is little time left to stop and think or to feel and be passionate. But it is passion that can save us and that spurs us onward to actions that surpass our own perceived limitations.

People meet in offices, bars, restaurants, grocery stores—in short, anywhere that other people congregate—but I met Nate at the top of a 200-foot redwood tree towering over the coastal hills of Humboldt County, in Northern California. Cristoforo, one of my old sea turtle project volunteers at Ría Lagartos, was a television producer in Italy. Through him, I landed a job as a television correspondent for one of the largest Italian television networks. My role was that of naturalist/expert/interviewer for an environmental program called *King Kong*.

My first assignment was to interview a tree-sitter. Nate was in his midtwenties, with long blond hair tied in a ponytail, green eyes, and an open and friendly smile. His physical appearance reminded me of one of those images of Jesus Christ one sees in churches, perhaps because of Nate's air of gentleness.

Eight months before we met, Nate left home one day for a walk among the ancient redwoods of Humboldt forest. As he hiked along the ridges, he passed the last remaining old-growth redwoods not yet felled by the Pacific Lumber conglomerate. He came into an open vista of previously dense forest, now slashed and burnt to nothing by the logging technique known as clear-cutting. The loggers were busily working away, and Nate was staggered by the horrifying wake of deforestation that stretched out before him. Without saying a word, he walked up to the last remaining sequoia in the tract and climbed up into its branches. He was determined to stop Pacific Lumber from felling this tree that had stood here for many generations. He chose to make his statement, and he did it without thinking twice.

The following morning, Nate's girlfriend found him perched like an owl, as she searched the trail, calling his name. She informed Nate's friends, who helped arrange for hoisting provisions and winter clothing up the tree to support Nate's effort, in between visits from the local sheriffs and patrols by the lumber-company security guards. As months passed, Nate became one with his tree, and his friends watched as he'd swing from branch to branch. He spent his time writing letters to newspapers about the plight of the redwoods while crouched under a tarp to avoid the rain or resting under the stars in the light of a full moon.

It was pouring rain on the day of the interview. As if it weren't nerve-racking enough to climb that colossal tree in a deluge—never having climbed anything before—that morning I learned the show would be broadcast live via satellite! From the base of the redwood, I could barely see Nate sitting on a branch with his legs hanging over as though it was the most natural thing in the world. I paused halfway up the tree to glance down at the television crew that seemed like a cluster of multihued mushrooms at the base.

The cameraman who came up after me was so scared that he was barely able to keep the camera still. As I sat with Nate near the top of that amazing old sequoia, stroked by wind and rain, I fully understood why he was there. He was driven by passion and respect for wild places. I liked Nate. To many he may have seemed naïve, but in Nate I recognized a true and innocent representation of love for nature.

From the forests to the oceans to the expansive deserts and beyond, we starve for more Nates in this world. We need to shake off our apathy and be passionate about our surroundings. It is a responsibility we all share, like it or not.

Captain Paul Watson is head of the world's most active and controversial environmental nonprofit organization: Sea Shepherd Conservation Society. Some folks love Watson; others hate him. Some think he's a nutcase, and others call him a hero. But whatever one might believe, this man has spent over thirty years passionately defending whales. He didn't just talk about the problems: he immersed himself in fixing them, directly and often at great risk. His efforts helped keep the issue of whaling in the public eye, in the face of massive and expensive attempts by whaling nations to sweep it under the table. I'm not suggesting that all of us become Nates and Paul Watsons, but I do believe that we must all, in our own ways, become personally involved in making our world better, even if it means pushing past our respective comfort zones.

On those dark occasions when I find myself wondering if anything I do really makes a difference, I need only think of the people I have seen become actively involved in protecting animals in the wild. Over the years, many of my volunteers have changed their careers and become, in one way or another, defenders of wildlife. This happened not because of me, but because of the firsthand experiences with nature they had while working with me. My programs and projects on the remote beaches of El Palmar, with whales in the Caribbean, and with dolphins in the waters of Greece and California were the means that brought them close to the natural grandeur they didn't know before.

Of the people I remember, there is Monica, who now heads a nonprofit organization focused on research and action-oriented conservation of sea turtles in Thailand, and Paul (now Pablo), who leads a manta conservation project in Mexico, and Mike, who is finishing up his PhD in oceanography with a strong and dedicated bias toward the protection of marine life. And there are many others who have changed their way of thinking and acting in their day-to-day life as a direct result of the time they spent in contact with wilderness and the animals that inhabit it.

Being outdoors enhances our creativity and ability to think critically, things that are slipping away from us in modern societies. We pass our days fragmented and interrupted by Facebook, Twitter, blogs, iPhones, in a digital microcosm that continuously ruffles our train of thought.

To protect dolphins, we need to protect oceans, and to do so, we must first consider ourselves an integral part of a grand natural system. Education, and more specifically environmental education, can help us to attain that goal. It can help us to better understand what we protect and why it is important to preserve it. Environmentalists are often seen as a fringe group of tree-hugging fanatics. But we are all environmentalists by the very fact that we all share this Earth on which we depend. Right wing, left wing, Catholic, Buddhist, housewife, businessperson, scientist, lawyer, or whatever . . . it doesn't matter.

"Environmentalism is not an option like choosing one's religion or political affiliation," my husband recently wrote in his book on the failure of environmental education. Charlie, like me, grew up with a passion for nature that changed later in life into a commitment to act in protection of vanishing wildness.

In Las Coloradas, it was hard to explain to the locals the significance of saving endangered sea turtles because conservation was outside their frame of reference. So where there was need, we had to find new solutions that made sense to the people living in those areas. In Trinidad, I witnessed firsthand how the work of my friend and colleague Scott Eckert has changed the fate of sea turtles on Matura Beach. Here, on one of the most significant nesting beaches for endangered leatherbacks, over two thousand females come every year to lay their eggs. In this place where sea turtle populations were declining at a daunting rate, I witnessed how Scott patiently provided the local communities with tools and impetus for the protection, management, and conservation of these animals. Using education, tolerance, and passion, Scott helped the locals move from poaching and exploitation to well-managed ecotourism programs that afforded the community a viable alternative form of income.

I don't want to seem overly simplistic. To save the oceans and its inhabitants, we'll need more than just passion. We already have sound science on our side, but we need to carefully address serious disconnects between the data provided by research and the creation and implementation of tangible, measurable, conservation-oriented action. We'll need the efforts of scientists, policy makers, economists, sociologists, and politicians working together.

We already have many practical solutions at our fingertips, such as marine policies that include all-inclusive ecosystem-based management approaches and establishment of marine protected areas in critical locations. We have some sensible ideas for building sustainable societies, such as the precautionary principle (which essentially states that great caution must be taken when doubt exists about potential adverse environmental impacts), the humility principle (which tells us to recognize our own limitations), and the reversibility principle (which advises against making irreversible changes). All good ideas to be sure, but most of our leaders haven't exactly proven themselves to be acting on behalf of the well-being of the planet.

I think that without the active involvement of a global community, we are not even coming close to making the changes necessary to mitigate the environmental problems we have created. Only through the individual cultivation of hope, love, passion, compassion, humility, education, and active involvement can we build a strong sense of stewardship of our world within our communities. People, and not only governments, must shape the direction in which decisions are made. Each of us has a say in the future of the dolphins, of whales, of the oceans, and of our own species. Exercising that say is our chance for greatness, perhaps the last chance.

In the end, it may all boil down to a question of morality. We are a moral species, or so we claim. But are our moral actions constrained to humans only? Historically, we have not extended our concepts of morality to cover the animal world because we see ourselves as the masters of our universe. But how's that working out for us so far? Judging from the mess we've made of the earth, not too well. We humans, who sit at the pinnacle of intelligence, have a moral responsibility to be kind to all things and beings, because we can.

If some charismatic animals can plead the case for changing our human attitudes toward nature and the protection of marine biodiversity, then dolphins and whales are surely prime candidates for the job. As for me, I continue to add to my mental library of experiences, surrounded by the immense beauty of the ocean wilderness and frequently in the company of wild dolphins. I have found myself in this life I have made, connected, as we are, to all living things on Earth. Narrating this journey is simply another attempt to reach out to the hearts and minds of others, so that we will find the way to protect and cherish the remarkable biodiversity that sustains us all.

List of Cetaceans Cited

Mysticeti (baleen whales)	
Family Balaenidae	
Southern right whale	*Eubalaena australis*
Family Eschrichtiidae	
Gray whale	*Eschrichtius robustus*
Family Balaenopteridae	
Common minke whale	*Balaenoptera acutorostrata*
Blue whale	*Balaenoptera musculus*
Fin whale	*Balaenoptera physalus*
Humpback whale	*Megaptera novaeangliae*
Odontoceti (toothed whales)	
Family Physeteridae	
Sperm whale	*Physeter macrocephalus*
Family Monodontidae	
White whale (beluga)	*Delphinapterus leucas*
Narwhal	*Monodon monoceros*

Family Phocoenidae	
Dall's porpoise	*Phocoenoides dalli*
Family Delphinidae	
Pacific white-sided dolphin	*Lagenorhynchus obliquidens*
Risso's dolphin	*Grampus griseus*
Common bottlenose dolphin	*Tursiops truncatus*
Spinner dolphin	*Stenella longirostris*
Pantropical spotted dolphin	*Stenella attenuata*
Striped dolphin	*Stenella coeruleoalba*
Short-beaked common dolphin	*Delphinus delphis*
Long-beaked common dolphin	*Delphinus capensis*
Killer whale (orca)	*Orcinus orca*
Short-finned pilot whale	*Globicephala macrorhynchos*
Family Ziphiidae	
Cuvier's beaked whale	*Ziphius cavirostris*
Family Lipotidae	
Baiji	*Lipotes vexillifer*

Suggested Readings

Bearzi, Maddalena, and Craig B. Stanford. *Beautiful Minds: The Parallel Lives of Great Apes and Dolphins*. Cambridge, MA: Harvard University Press, 2008.

Bekoff, Marc. *The Emotional Lives of Animals*. Novato, CA: New World Library, 2007.

———. *Minding Animals: Awareness, Emotions, and Heart*. Oxford: Oxford University Press, 2002.

Bekoff, Marc, and Jessica Pierce. *Wild Justice: The Moral Lives of Animals*. Chicago: University of Chicago Press, 2009.

Berta, Annalisa, and James L. Sumich. *Marine Mammals: Evolutionary Biology*. San Diego: Academic Press, 1999.

Boyd, Ian L., W. Don Bowen, and Sara J. Iverson. *Marine Mammal Ecology and Conservation: A Handbook of Techniques*. London: Oxford University Press, 2010.

Byrne, Richard W., and Andrew Whiten, eds. *Machiavellian Intelligence*. Oxford: Clarendon Press, 1988.

Carson, Rachel L. *The Sea around Us*. New York: Oxford University Press, 1989.

———. *Silent Spring*. Boston: Houghton Mifflin Co., 1994.

Chatwin, Bruce. *In Patagonia*. New York: Penguin Books, 1977.

Connor, Richard C., and Dawn M. Peterson. *The Lives of Whales and Dolphins*. New York: Henry Holt and Co., 1994.

Cousteau, Jacques, and Susan Shiefelbein. *The Human, the Orchid, and the Octopus: Exploring and Conserving Our Natural World*. New York: Bloomsbury, 2007.

Dailey, Murray D., Donald J. Reish, and Jack W. Anderson, eds. *Ecology of the Southern California Bight: A Synthesis and Interpretation*. Berkeley: University of California Press, 1993.

Darwin, Charles. *The Expression of the Emotions in Man and Animals*. 3rd edition. Oxford: Oxford University Press, 1998.

———. *On the Origin of Species*. 1859. Reprint, London: Penguin Classic, 2009.

Devine, Eleanore, and Martha Clark, eds. *The Dolphin Smile*. New York: Macmillan Co., 1967.

De Waal, Frans B. M. *The Age of Empathy: Nature's Lessons for a Kinder Society*. New York: Harmony Books, 2009.

De Waal, Frans B. M., and Peter L. Tyack, eds. *Animal Social Complexity: Intelligence, Culture, and Individualized Societies*. Cambridge, MA: Harvard University Press, 2003.

Dewey, John. *Democracy and Education*. Middlesex, England: Echo Library, 2007.

Ellis, Richard. *The Empty Ocean*. Washington, DC: Island Press, 2003.

———. *Tuna: A Love Story*. New York: Alfred A. Knopf, 2008.

Frohoff, Toni, and Brenda Peterson, eds. *Between Species: Celebrating the Dolphin-Human Bond*. San Francisco: Sierra Club Books, 2003.

Goodall, Jane. *The Chimpanzees of Gombe: Patterns of Behavior*. Cambridge, MA: Harvard University Press, 1986.

Harcourt, Alexander H., and Frans B. M. de Waal, eds. *Coalitions and Alliances in Humans and Other Animals*. Oxford: Oxford University Press, 1992.

Hauser, Mark D. *Wild Minds*. New York: Henry Holt and Co., 2000.

Herman, Luis M. *Cetacean Behavior: Mechanisms and Functions*. New York: Krieger Publishing Co., 1988.

Heyes, Cecilia, and Bennett Galef, Jr., eds. *Social Learning in Animals: The Roots of Culture*. New York: Academic Press, 1996.

Heyning, John. *Whales, Dolphins, Porpoises: Masters of the Ocean Realm*. Seattle: University of Washington Press, 1995.

Leatherwood, Stephen, and Randall R. Reeves. *The Bottlenose Dolphin*. San Diego: Academic Press, 1990.

———. *The Sierra Club Handbook of Whales and Dolphins*. San Francisco: Sierra Club Books, 1983.

Leopold, Aldo. *A Sand County Almanac, and Sketches Here and There*. New York: Oxford University Press, 1948.

Lorenz, Konrad. *King Solomon's Ring: New Light on Animal Ways*. 2nd edition. New York: Routledge, 2002.

Louv, Richard. *Last Child in the Woods: Saving Our Children from Nature-Deficit Disorder*. Chapel Hill, NC: Algonquin Books, 2008.

Mann, Janet, Richard Connor, Peter L. Tyack, and Hal Whitehead, eds. *Cetacean Societies: Field Studies of Whales and Dolphins*. Chicago: University of Chicago Press, 2000.

Masson, Jeffrey M., and Susan McCarthy. *When Elephants Weep: The Emotional Lives of Animals*. New York: Delta Book, 1995.

Meffe, Gary K., and Ronald C. Carroll. *Principles of Conservation Biology*. 3rd edition. Sunderland, MA: Sinauer Associates, 2006.

Melville, Herman. *Moby Dick; or, The Whale*. 150th anniversary edition. New York: Penguin Books, 2001.

Norris, Ken S. *The Porpoise Watcher*. New York: W. W. Norton and Co., 1974.

Norris, Ken S., Bernd Würsig, Randall S. Wells, and Melany Würsig. *The Hawaiian Spinner Dolphin*. Berkeley: University of California Press, 1994.

Norse, Elliott A., and Larry B. Crowder. *Marine Conservation Biology: The Science of Maintaining the Seas' Biodiversity*. Washington, DC: Island Press, 2005.

Orr, David. *Earth in Mind: On Education, Environment and the Human Prospect*. Washington, DC: Island Press, 1994.
Payne, Roger. *Among Whales*. New York: Delta, 1995.
Perrin, William F., Bernd Würsig, and J. G. M. Thewissen, eds. *Encyclopedia of Marine Mammals*. San Diego: Academic Press, 2002.
Postman, Neil. *The End of Education: Redefining the Value of School*. New York: Vintage Books, 1995.
Pryor, Karen, and Ken S. Norris. *Dolphin Societies: Discoveries and Puzzles*. Berkeley: University of California Press, 1991.
Reynolds, John E., Randall S. Wells, and Samantha D. Eide. *The Bottlenose Dolphin: Biology and Conservation*. Gainesville: University Press of Florida, 2000.
Reynolds, John E., William F. Perrin, Randall R. Reeves, Suzanne Montgomery, and Timothy J. Ragen, eds. *Marine Mammal Research: Conservation beyond Crisis*. Baltimore: Johns Hopkins University Press, 2005.
Reynolds, John E., and Sentiel A. Rommel, eds. *Biology of Marine Mammals*. Washington, DC: Smithsonian Institution Press, 1999.
Riedmann, Marianne. *The Pinnipeds: Seals, Sea Lions, and Walruses*. Berkeley: University of California Press, 1990.
Safina, Carl. *Song for the Blue Ocean*. New York: Owl Books, 1997.
———. *Voyage of the Turtle: In Pursuit of the Earth's Last Dinosaur*. New York: Henry Holt and Co., 2006.
Saylan, Charles, and Daniel T. Blumstein. *The Failure of Environmental Education (and How We Can Fix It)*. Berkeley: University of California Press, 2011.
Sepúlveda, Luis. *Patagonia Express (Fábula)*. Spanish edition. Barcelona: Tusquets, 2001.
Singer, Peter. *Animal Liberation*. Revised edition. New York: HarperCollins Publishers, 2002.
Sobel, David. *Place-Based Education: Connecting Classrooms & Communities*. Great Barrington, MA: Orion Society, 2005.
Stanford, Craig B. *Chimpanzee and Red Colobus: The Ecology of Predator and Prey*. Cambridge, MA: Harvard University Press, 1998.
———. *Significant Others: The Ape-Human Continuum and the Quest for Human Nature*. New York: Basic Books, 2001.
Thoreau, Henry D. *Walden and Other Stories*. New York: Modern Library Classic, 2000.
Tinbergen, Niko. *Curious Naturalists*. Revised edition. Amherst: University of Massachusetts Press, 1984.
Twiss, John R., Jr., and Randall R. Reeves, eds. *Conservation and Management of Marine Mammals*. Washington, DC: Smithsonian Institution Press, 1999.
Wasserman, Edward A., and Thomas R. Zentall, eds. *Comparative Cognition: Experimental Explorations of Animal Intelligence*. Oxford: Oxford University Press, 2006.
White, Thomas I. *In Defense of Dolphins: The New Moral Frontier*. Malden, MA: Blackwell Publishing, 2007.
Wilson, Edward O. *Biophilia*. Cambridge, MA: Harvard University Press, 1984.
———. *Sociobiology: The New Synthesis*. Cambridge, MA: Belknap, 1975.
Woodard, Colin. *Ocean's End: Travels through Endangered Seas*. New York: Basic Books, 2000.

Index